Praise for David Jenyns & Systems Champion

"Systems Champion delivers exactly what every business owner needs. David Jenyns has created a practical framework that removes the owner from systems creation process and puts that responsibility where it belongs – with a dedicated function called, "Systems Champion". This book will revolutionize how you think about and implement systems in your business."

— **Gino Wickman, author of *Traction* and *Shine*, Creator of EOS®**

"The Systems Champion Academy is essential for anyone who owns a business today."

— **Michael E. Gerber, author of *The E-Myth***

"This book is the missing manual for entrepreneurs who want a scalable, stable and systemised business."

— **Allan Dib, author of *The 1-Page Marketing Plan***

"Systems aren't sexy. But freedom is. This book gives you the step-by-step to get both."

— **Mike Michalowicz, author of *Clockwork* and *Profit First***

"Systems Champion breaks the biggest myth in business that systemizing is the leader's job alone. David Jenyns shows that true leadership lies in empowering others to build systems that drive the business forward, without constant owner involvement."

— **Brad Sugars, founder of ActionCOACH**

"What I love about David's approach is how genuinely he cares about helping you succeed. He doesn't just dump theory on you - he walks you through real stories of business owners who were drowning in daily operations and gives you the 'secret sauce' to make a real change."

— **Kerry Boulton, founder of The Exit Strategy Group**

"Systems Champion isn't just another business book, it's a practical playbook for anyone who wants to make meaningful change inside a company. If you're serious about creating more career success, this is essential reading."

— **Dale Beaumont, founder and CEO of Business Blueprint®**

"Take your life back from your business, starting now! This book is your blueprint for reclaiming time, focus, and freedom. You'll wonder why you didn't do it sooner. Cheers!"

— **Dr Sabrina Starling, author of *The 4 Week Vacation*®**

"I've known Dave for fifteen years, watching our professional relationship grow into one of my most valued friendships. What sets Dave apart is how deeply he engages with everything he does. He's a genuine advocate for small businesses who brings both expertise and heart to his work."

— **Mike Rhodes, founder of 8020agent.com**

"I've known Dave for nearly a decade, and he's one of the most generous, committed, and value-driven people I've met. Dave walks his talk, delivers beyond what he promises, and leaves people better off than before they met him. The quality of personal and business connections he has speaks volumes about the quality of his character and values."

— **Simon Kelly, CEO of Seriously Good Design**

SYSTEMS CHAMPION

Simplify Business Processes,
Unlock Team Potential &
Achieve True Freedom

By **David Jenyns**

Published by SYSTEMology
www.SYSTEMology.com

The moral right of the author has been asserted.

A catalogue record for this book is available from the National Library of Australia.

ISBN: 978-0-6488710-4-0 (paperback)
ISBN: 978-0-6488710-7-1 (hardback)
ISBN: 978-0-6488710-5-7 (ebook)
ISBN: 978-0-6488710-6-4 (audiobook)

Books by the author:

Authority Content: The Simple System for Building Your Brand, Sales, and Credibility (2016)

SYSTEMology: Create Time, Reduce Errors and Scale Your Profits with Proven Business Systems (2020)

Systems Champion: Simplify Business Processes, Unlock Team Potential and Achieve True Freedom (2025)

For quantity sales or media enquiries, please contact the publisher at
www.SystemsChampion.com

Editing by Kelsey Garlick
Cover Design by July Aarillo
Interior Layout by Olivier Darbonville
Publishing Consultant: Linda Diggle

Disclaimer: Although the author and publisher have made every effort to ensure the information in this book was correct at press time, the author and publisher do not assume and hereby disclaim any liability to any party for any loss, damage, or disruption caused by errors or omissions, whether such errors or omissions result from negligence, accident, or any other cause.

To Carrolyn, Nate and Jordan.
Thank you for always reminding me to drink the good wine.

Contents

PILLAR 3: CULTURE

IMPLEMENTATION

Definition

Sys-tems: A structured set of processes or practices that work together to achieve a specific outcome.

Cham-pi-on: A dedicated individual who takes ownership of a cause, consistently advocating for its importance and driving its success.

Important note: Throughout this book, I've used terms such as "systems", "processes", "how-to documents", "standard operating procedures (SOP)", "workflows", etc. somewhat interchangeably. While there are differences in their technical meanings, I believe keeping things simple is more important as you're starting out on your systemisation journey. The focus should be on creating consistency in your business operations, not getting caught up in terminology.

Introduction

I HAD BEEN DREADING THIS VIDEO call for a month. I felt sick. My heart was racing as I clicked the meeting link, a heaviness in my chest that had been growing since I first realised what needed to be done. This was the last of seven team members I had to let go, all within a couple of months – a decision that I did not take lightly. It was something I had been thinking about and doing my best to avoid over the better part of the year.

I'd deliberately postponed this call until after Christmas, thinking I was doing her a favour by not ruining her holidays. Now, staring at her name in the virtual waiting room, I wasn't so sure.

She had been with me for more than five years. I knew her son, her dreams of buying a house, how passionate she was about the work we did together. This wasn't just business. It was personal. It always is when you build a team the right way.

As her face appeared on screen, memories of our team gathering in the Philippines flooded back. We'd rented a sprawling house on top of a mountain with enough rooms for everyone. My CEO was American, and she had cooked a full Thanksgiving dinner with all the trimmings. We'd gathered around the long dining table, spilling out onto the deck, sharing stories and laughter. We had so much fun. The night turned to karaoke, where our Filipino team members (and some of the Aussies) took to the microphone with so much enthusiasm, belting out songs until one in the morning. It hadn't felt like work; it had felt like family.

When our eyes met through the screen, I fought to keep my voice steady. As I delivered the news, I had to mute myself for a moment to compose myself … to fight back the tears. There was a long silence on the other end.

"I wish you had told me before Christmas," she finally said, her voice smaller than I'd ever heard it. "I would have had more time to prepare."

Her words hit me like a punch to the guts. I'd convinced myself I was protecting her, but in reality, I'd only been protecting myself, delaying the inevitable, robbing her of precious time to prepare.

"I'm sorry," I said with a lump in my throat.

I still check in with some of the team members we had to let go. I helped place several in other organisations, and through my connections, I get updates about how they're going. Most have landed on their feet, finding new roles that, in some cases, have pushed them to grow in ways they might never have with us. It's a small consolation, but a consolation nonetheless.

The reality was, this wasn't just one tough decision but the culmination of a technological shift that had forced my hand. We had built a thriving business helping companies implement systems, training professionals (called SYSTEMologists) who would transform businesses through documented processes. One challenge we consistently solved was the time-consuming task of turning video recordings of tasks being completed into clear, actionable systems documentation. Our team excelled at taking raw materials and crafting them into valuable business assets.

Then artificial intelligence (AI) changed the game virtually overnight.

It started with ChatGPT, and then other tools emerged that could accomplish in minutes what had previously taken our team hours. What we had built as a competitive advantage suddenly became accessible to anyone. Within a few months, demand for our documentation services dried up. The value we provided hadn't diminished, but the way it needed to be delivered had fundamentally changed.

I tried relocating the documentation team to different roles. I even gave one team member a range of my own personal tasks to keep them busy. But the reality was unavoidable: we were suddenly way overstaffed, and I knew we had to take more drastic action.

Alone in my office after letting that last team member go, I was pretty emotional. Letting go of staff is never easy. I questioned everything. What else was about to change? What other foundations of our business model might suddenly shift? In a world moving this quickly, how could anyone build something stable and lasting?

I learned a long time ago that panic never solves problems. I took a deep breath and forced myself to think clearly. The conclusion wasn't comfortable, but it was necessary. Now is the time for action. Not next quarter. Not when things settle down. Now.

This isn't about fear. It's about seeing reality clearly and making tough decisions that protect the greater good of your business. A business that doesn't adapt doesn't just fail its owner, it fails everyone who depends on it.

The most valuable insight came from an unexpected place. As our team adapted to new technologies, I realised something counterintuitive: processes weren't becoming less important – they were becoming more important than ever!

To get great results with AI, we needed clear instructions, well-defined outcomes and structured data. The clearer our process, the better our results. Process is the programming for the machines. This realisation got my team and I to double down on our systems-first approach.

Together we proved strong systems, powered by AI, deliver exceptional results. Tasks that once took days now took hours, projects that seemed impossible became routine, and we found ourselves delivering more value than ever before. The painful layoffs became the catalyst for reinvention. Within 12 months, our business had not only recovered but was growing again.

My story isn't unique. Business owners everywhere are facing similar moments of truth, forced to make difficult decisions as technology reshapes what's possible. I don't share this with you to cause alarm, but out of genuine care and concern for your future. The business landscape is transforming at an unprecedented pace, and those who adapt will thrive while those who resist may find themselves left behind.

Intuitively, you and I both know systems-driven businesses are better businesses. They're better for clients, for staff, for owners and for the bottom line. They're more organised, adapt more easily and allow businesses to reach their true potential.

There's no debate as to why you'd want to own or be a part of a systemised business.

But why has it historically been so hard? Why do so many companies struggle to make it work? Is it time or resources? Is it that they don't know where to start? Or perhaps they simply can't get the team on board?

In truth, there are an infinite number of reasons why people fail to make it work and the sad reality is, the vast majority of small businesses never reach the promised land of systemisation bliss.

The good news is, it doesn't have to be this way.

When I wrote *SYSTEMology* (my last book) a few years back, I had a dream: to free business owners worldwide from the day-to-day operations of running their businesses. That book helped thousands of businesses take their first steps toward systemisation, and I'm humbled by the transformations I've witnessed.

But the world has changed dramatically since then. This technological revolution has created an extraordinary opportunity and the gap between systemised and unsystemised businesses isn't just widening – it's accelerating!

But here's the uncomfortable truth I've discovered through working with thousands of businesses. Technology alone isn't enough. AI tools, no matter how powerful, can't drive transformation by themselves. They need someone to harness their potential, to bridge the gap between possibility and reality.

This is where the Systems Champion holds the key.

The businesses that adapt and thrive aren't just the ones with the best tools or most innovative ideas. They're the ones who have someone dedicated to make this transformation happen. A Systems Champion who learns and masters both the technical and human elements, steadily turning vision into reality.

The Systems Champion isn't just another administrative role. They keep systems front and centre, constantly making progress when everyone else is caught up in the daily whirlwind.

It's this consistent progress that makes all the difference. A documented process here, a refined workflow there, a new automation that saves everyone 10 minutes. Together they create a compound effect that transforms your business one system at a time.

But it's more than just improving the processes. Systems Champions become cultural catalysts for lasting change. While the business owner focuses on growth, Systems Champions build the foundations to ensure everything runs efficiently. They make systems part of your company's DNA, setting standards and making it easy for everyone to capture and share their knowledge.

If selected wisely, the Systems Champion may even become arguably one of the most important team members in the organisation. Some have the potential to evolve into exceptional operations managers because they gain an unparalleled view of how the business runs.

Long story short, this book represents a fundamental shift in how we approach business systemisation. While *SYSTEMology* was written for business owners, *Systems Champion* speaks to both owners and the champions who will drive this transformation forward.

The early chapters will help business owners identify and empower the right person for this crucial role. The heart of the book then shifts to become a practical playbook for these champions, giving them everything they will need from documentation strategies to team adoption techniques.

It's an exciting time! You're about to embark on an extraordinary journey. Whether you're a business owner ready to empower your Systems Champion or a champion preparing to take on this crucial role, this book will show you how to harness the power of systems to transform your business.

The future belongs to businesses that can systemise and adapt effectively in this new world. The tools are here. The opportunity is now. All that's missing is the right person to champion the cause.

Let's begin.

David Jenyns
Founder, SYSTEMology

FOR THE
BUSINESS OWNER

Summary

Most business owners know systems-driven businesses are better businesses, yet systemisation efforts often fail. The missing piece? Having the right person driving the implementation forward – your Systems Champion.

A dedicated Systems Champion transforms how businesses approach systemisation, enabling owners to focus on vision and growth while ensuring systems become part of the company's DNA.

Highlights covered in these chapters include:

- The critical distinction between being a champion of systems versus being the Systems Champion.

- How finding the right Systems Champion is the missing link in successful business systemisation.

- Why business owners are often the bottleneck in systemisation despite their best intentions.

- The eight essential qualities that make an effective Systems Champion, from organisational prowess to leadership potential.

- The emerging connection between systems thinking and AI implementation, and how Systems Champions naturally evolve into AI Champions.

- Why proper support and empowerment of your Systems Champion is crucial for business transformation.

1

This Book Is Not for You

YOU WANT TO BUILD A systems-driven business. You've known for a while that systemisation[1] is the key to growth, efficiency and freedom from day-to-day operations. You've probably even tried to make it happen.

And yet ...

Here you are, still caught in the chaos of day-to-day business operations. Still putting out fires. Still feeling like the business depends too much on you.

I get it. I've seen this pattern hundreds of times. Despite your best intentions and countless attempts, systemisation remains just out of reach. The systems you document gather digital dust. The standards you create get ignored or forgotten. The tools you implement don't get used.

But all is not lost. First, let me acknowledge something important. The fact that you're reading this book puts you ahead of most business owners. You understand that systems are the key to scaling your business and creating true freedom. That's no small thing.

---

1   The word "systemise" has many definitions, but in the context of SYSTEMology it means the documentation of a system (or series of systems) within a business, so that it can be repeatedly replicated to the same standard, either by a human or by technology.

You're thinking beyond the immediate challenges. Beyond the daily fires that need extinguishing. You're thinking about building something that lasts. Something that can run without your constant attention.

You started your business with a passion for solving client problems. But I'm willing to bet you also dreamed of freedom. The freedom to take a vacation without checking your phone every hour. The freedom to focus on growth rather than operations. The freedom to choose how you spend your time.

And you're absolutely right. Systems are the key to that freedom.

You're just one missing piece away from making systemisation work in your business. In my years of working with business owners, I've seen the same challenges appear over and over:

- "I don't have enough time to work on systems."
- "I can't figure out where to start."
- "My team won't follow the processes."
- "I get lost in the details."
- "The tools are overwhelming."

But no matter the challenge, they all share the same solution. It's not about finding more time, or the perfect tool, or a better process. It's about recognising one fundamental truth:

> You are not the right person to champion systems in your business.

This isn't a criticism. It's liberation.

As a business owner, your superpower is vision. You see opportunities others miss. You think in possibilities and potential. You're built to spot the next big move, not document the current one.

Your natural tendency is to look forward, to strategise, to grow. And that's exactly as it should be! Your business needs that visionary energy. What it doesn't need is for you to get bogged down in the details of process documentation.

This is why your previous attempts at systemisation might have stalled. It's not because you lack capability or commitment. It's because you're trying to be something you're not.

This is why this book isn't for you – at least, not most of it. It's for your Systems Champion.

The first few chapters will show you how to find and empower the right person for this role. We'll cover the qualities to look for, how to set your champion up for success and how to support their work without micromanaging.

But after that? This book belongs to your Systems Champion. It's their manual, their playbook, their guide to making your systems vision a reality.

Your role is to pass this book to the right person and give them the authority to run with it. By all means, read the whole thing. You should know what your Systems Champion will be doing. But resist the urge to implement it yourself.

Before you turn another page, make a commitment to yourself. Make a commitment to yourself, your health, your wellbeing and to those you truly care about. Commit to letting go of being the systems person in your business. Commit to finding and empowering someone else to take on this crucial role.

This might feel uncomfortable. After all, systems are vital to your business's success. How can you step back from something so important?

But here's the truth: stepping back is the most important thing you can do. Your business will make more progress toward systemisation when you stop trying to drive it yourself and instead empower the right person to champion the cause.

Your dreams of a business that runs without you? They're closer than you think. But first, you need to accept that this book isn't for you. It's for your Systems Champion.

Are you ready to make that shift?

Let's find your Systems Champion.

Your Liberation Plan

I commit to:

☑   Stop being the systems person in my business

❑   Identify a Systems Champion by (date): ....

❑   Hand over systems responsibility by (date): ....

❑   Support but not control the systems process

❑   Provide necessary resources and authority

Signed: ....

Date: ....

2

Finding Your Systems Champion

Y OU MAY BE A CHAMPION of systems, but the Systems Champion you hire (or promote from within) is going to be the person that drives your SYSTEMology project forward. The skills required to be effective in this role are specific, yet interestingly, prior experience in this kind of work isn't necessarily needed. The Systems Champion role is one that can be learned and mastered. But like any dedicated task, some people are better suited to it than others.

Before we dig into the ideal qualities required for this challenge, let's get specific for a moment. What exactly is a Systems Champion and what are their responsibilities? Since this is the core theme of this book, it's important that we're both on the same page.

First, let's be crystal clear about what a Systems Champion is NOT. They're not your chief operating officer (COO), second-in-command, operations manager, or Integrator. The Systems Champion reports TO these leadership roles (or the business owner in smaller teams) and they don't replace them.

Think of it this way. They are like the department head of the "systems" department, if "systems" was a department like sales, marketing, HR or finance. Just as your sales manager owns the sales function and your marketing manager owns marketing, your Systems Champion owns the systems function. And just like those other department heads, they report up to senior leadership while having authority within their domain.

So what ARE they? In brief, the Systems Champion is ***the person who manages and drives the systemisation of your business forward.*** They will physically take responsibility for documenting processes, organising them into systems and ensuring those systems are implemented and followed. Alongside the practical activities, the role also calls for the development of a systems-driven culture within the business. They'll be responsible for educating the rest of your team on what's required of them and encouraging them to support the systemisation project.

Needless to say, your Systems Champion plays a vital role in transitioning the business toward a more systemised and scalable operation.

That's the 30,000-foot view. But what do the responsibilities of your champion actually include? The following is just an overview. The section of this book written for your Systems Champion will cover all of this in much greater detail.

Defining systemisation needs: In collaboration with the business owner and team leaders, your Systems Champion will decide which processes are suitable for systemisation (tasks that are essential, recurring, already working in your business and suitable for delegation).

Assigning knowledgeable workers: Your Systems Champion will identify which personnel within the company have the most knowledge about specific tasks or processes ("knowledgeable workers"). They will determine who these individuals are through discussions with team members and department heads.

Extracting systems: By conducting interviews with the knowledgeable workers, your Systems Champion will extract their wisdom and experience and document it in a manner that can be understood easily by others in the business.

Documentation and formatting: Consistency of documentation is critical, which means the Systems Champion needs to define a unified style, layout and level of detail. This may include tasks such as transcribing recordings, formatting documents and incorporating visual aids.

Managing system storage and accessibility: Setting up systems management software, such as systemHUB, as your single source of truth is a critical responsibility for your Systems Champion. Your documentation should live in one central location that is logical and easily searchable, not scattered across shared drives, email attachments or personal computers.

Accountability and transparency: The Systems Champion will ensure systems are clearly visible and accessible, so that accountability comes naturally to your team. This includes helping to create an environment where doing the right thing is the easiest path forward and eliminating excuses even before they arise.

Systems adoption and culture: Fulfilling the "champion" element of their job title, your Systems Champion will communicate the importance of systemisation, get buy-in from the team and encourage everyone to follow documented processes. This might include individual and group training sessions, as well as the creation of incentive initiatives.

Some of the above responsibilities may be obvious to you and others not so. Most business owners, when they first review the duties held by a Systems Champion, are surprised by how complex and wide-ranging the role truly is. It requires strong attention to detail, excellent communication skills, general problem-solving abilities, an organised mindset and a proactive drive.

Your Systems Champion, whoever they turn out to be, is going to be an intelligent, flexible and energetic individual.

After talking to and observing countless successful Systems Champions, I've identified five key qualities that indicate their likely ability to manage this role. It isn't necessary for them to embody all of these qualities to a high degree, but they should have at least some level of proficiency in all of these areas.

1. Organisational skills and detail orientation: Order and structure are the foundation of SYSTEMology; therefore, a Systems Champion needs to be naturally organised with a sharp eye for detail. There isn't room for shortcutting in this role, so a meticulous nature is key.

2. Exceptional communication and interpersonal skills: Your Systems Champion will be liaising with everyone in your business at all levels. They must be as comfortable and confident in discussions with senior management as they are with the entry-level workers. Self-confidence and assertiveness must be part of this mix because there may be times when they need to exert some gentle authority.

3. Curiosity and creative problem-solving abilities: Roadblocks and obstacles are common to every SYSTEMology implementation. While some of them are common and are discussed in these pages, every company will have its own unique challenges that will require your Systems Champion to problem-solve creatively. Curiosity is also important because, when they are "extracting" information from the knowledgeable worker, they must seek to understand the processes involved and ask the right questions to draw out the details.

4. Adaptability and tech-savviness: Technology (including AI) is going to be used through the SYSTEMology process. Your Systems Champion needs to be comfortable learning and using new software, even connecting systems together. And while consistency is the goal, as your business and the technology that drives it changes, your champion needs to be comfortable adjusting with it.

5. Leadership potential and assertiveness: Your Systems Champion does not need to be an existing leader, or even to become one as part of this role, but they do need some of the qualities inherent to leaders. In fact, leadership team members often make poor Systems Champions as they're typically time-poor and pulled in many directions. Instead, look for someone who has capacity and can manage their own schedule and output while bringing others along with them. They need the confidence to convince those who might be reluctant to engage with the program (yes, this can happen, and we'll cover this in detail) but without necessarily holding a formal leadership position.

To echo Angelica in the musical *Hamilton*, you're "looking for a mind at work". This isn't a role for someone who is passive or most comfortable following instructions. There's an almost entrepreneurial quality to a good Systems Champion and a keen level of intelligence is a minimum requirement.

Does this mean you're looking for a unicorn? No. Finding someone who excels in all these areas would be fantastic, but it's not necessary. Look for a person who shows potential in these qualities – they can develop and strengthen their skills over time with the right training. What's most important is finding someone with genuine enthusiasm and willingness to take ownership of this project.

Do you have someone in mind?

The time commitment for a Systems Champion can vary significantly based on your business size, goals and available resources. At minimum, you should allocate half a day per week to get started. A part-time commitment will help you make steady progress, while a full-time Systems Champion can drive faster transformation.

For a small business under 10 team members, starting with a few dedicated hours each week might be sufficient, especially if you're assigning this responsibility to an existing team member with some bandwidth. As your business grows or if you want to accelerate your systems implementation, you may need to increase these hours. The key is to begin with a realistic time commitment that matches your business's current needs and resources, then adjust as needed.

Perhaps you can already think of someone within your business who is a perfect, or at least a strong, fit for this job. And it's great if you can find someone internally. They'll already have a head start because they will have an existing familiarity with your company and how it operates. An apprentice, for example, who is keen to learn and develop can be a great option because, presuming they have the right qualities, this is a great role for gaining a deeper understanding of how your business functions.

A returning-to-work mum or dad can also be a good option. Perhaps you have someone who used to work for you many years ago and is seeking a return to employment. This could be a great project to get them back into the workplace, with the added advantage that they likely already know many of your team and your departments.

But don't force it. If you can't find someone within your existing team who is a strong match for the above qualities, that's okay. You shouldn't settle for someone who isn't a good fit, just because you know them. In fact, bringing someone in externally, while they will need some time to get to grips with how your business operates, can have its own advantages. A fresh pair of eyes can be a valuable asset.

Hiring someone externally is going to require the same process for finding any new team members, but with the added wrinkle that most of the people who are going to be the right fit will probably have never heard of the role. This will hopefully change in the future, but at the time of writing, it's unlikely that many people are going to be putting "Systems Champion" into a job search engine.

No problem. I've already tackled the challenge of creating a suitable position description and job advert. You can find those below in appendix 1.0 and 1.1 or download them here:

www.SystemsChampion.com/resources

Have a few people already in mind? Compare them using the following evaluation matrix.

Systems Champion Evaluation Matrix

Rate potential candidates on each quality (1–5):

❑ Organisational skills   1   2   3   4   5

❑ Communication skills   1   2   3   4   5

❑ Problem-solving ability   1   2   3   4   5

❑ Tech-savviness   1   2   3   4   5

❑ Leadership potential   1   2   3   4   5

❑ Initiative   1   2   3   4   5

❑ Capacity   1   2   3   4   5

Compare multiple candidates:

1. .... Score: .... / 35

2. .... Score: .... / 35

3. .... Score: .... / 35

The AI Advantage

IS CREATING AN ENTIRELY NEW job position for your systemisation project excessive, or even a luxury? Well, consider that the alternative is to perform the role yourself – a role that is, for most businesses, a full-time position. Hopefully, by this stage, you recognise that this project is too important and too involved to be performed in the tiny pockets of spare time you have available alongside running your business.

And let's not forget that one of your goals with this process is to eventually remove yourself from the day-to-day running of your business. This isn't going to happen by giving yourself more work to do creating and managing systems. Ultimately, the cost of hiring a new team member (or moving an existing team member away from some or all of their current responsibilities) is a small price to pay to achieve the goal.

Keep in mind that as the power of systems takes hold, the improved efficiency your business enjoys will translate into additional profits, which will eventually more than cover the costs of a new hire. And while we're considering the long view, here's another major benefit of having a Systems Champion in your business …

A Systems Champion is uniquely positioned to also become your AI Champion.

Depending on when you read this, integrating AI into your workflows may not be on your radar just yet. But it's only a matter of time before you either need to consider it to stay at the forefront of your industry, or it will naturally find its way in through upgrades to software and systems that you're already using.

AI has already moved past the period of gimmick and has become a core tool for many professions. SYSTEMology is one of them.

As you'll see, AI is a valuable tool with which your Systems Champion is going to become very familiar. But this is only the beginning. As AI starts to creep its way into more areas of your business, you'll need someone on your team who understands its benefit, can help your departments implement it effectively and also become a champion for its use among your team, who may still be suspicious or nervous about the technology.

But your Systems Champion eventually becoming your AI Champion is more than just convenience. There's a natural overlap in the skills required, and their eventual mastery of systems is going to give them a natural affinity for AI.

If you've played around with any general AI tools, such as ChatGPT or Gemini, you'll likely have noticed that much of AI is built around "prompts"[2]. Getting AI to perform optimally and produce awesome results is about understanding how to construct prompts in a way that

---

2   In the context of AI, a "prompt" is set of instructions that a human gives to an AI system to achieve a desired outcome.

the technology can understand. Some savvy individuals are giving themselves the title of prompt engineer, and there are already training courses in the skill, and pre-written prompt libraries, available for sale.

This is really interesting to me because a prompt is simply a system that you're using to instruct the AI. And a system, in the world of SYSTEMology, is a series of steps that when followed create a consistent outcome. Can you see the connection?

As a Systems Champion develops their skills in identifying, documenting and optimising processes, they're also becoming familiar with the language of systems, including the details and the logical flow. This translates perfectly to prompt engineering. If someone can break down a complex process into a clear, replicable set of steps that humans can follow, it's only a small lateral step into doing the same in a way that AI can follow.

This is only the beginning of the overlap. Your Systems Champion is going to become very familiar with your business's pain points and the areas in which efficiency gains would have the biggest impact. This is exactly the knowledge that is required to identify areas in which AI can help.

Last, but not least, a great Systems Champion is a natural optimiser. Over time they become attuned to the "music" of systems, and when they spot a step that is unnecessary or that can be streamlined, it jumps out at them like a singer hitting a wrong note. Their natural drive for continuous improvement will lead them right into the world of AI opportunities that are available, and they'll soon start coming to you with recommendations for areas in which AI should be considered.

Like it or not, the companies that will lead the way in their industries over the next few decades will be the ones that integrate AI. Not as a distraction, or as a PR stunt, but as a genuine means for creating a more optimised business that can reduce costs, speed up work processes and deliver additional value to the customer. As AI technology continues evolving, the role of AI Champion is going to become increasingly popular and valuable across all industries. Hiring a Systems and AI Champion now is essentially futureproofing your business – you'll have someone who is naturally immersed in the technology and can help you use it to best effect sooner rather than later.

These are all points to keep in mind when seeking your Systems Champion. What I really want you to think about here is that this is a valuable role for a whole range of reasons. Don't make the mistake of thinking of it as a glorified administrative role or a project that just anyone can take on and make a success of. This is an important role that can transform your business over the next few years, so treat the hiring process with the level of care and consideration it deserves.

AI Readiness Assessment

How is your business approaching AI today?

Ignoring it → Exploring options → Actively implementing

Rating 1–10: ....

1. What AI tools are you currently using?

❑ None

❑ Basic (e.g. existing tools, chatbots)

❑ Intermediate (e.g. AI automation tools)

❑ Advanced (e.g. custom-coded AI solutions)

2. Do you know of any quick wins with AI?

❑ Customer service

❑ Content creation

❑ Data analysis

❑ Process automation

❑ Other: ....

3. What's your biggest concern about AI?

❑ Uncertainty about where to start

❑ Potential business disruption

❑ Implementation costs

❑ Team adoption challenges

❑ Other: ....

4

This Is Your Stop

A SYSTEM FOR SYSTEMISING YOUR BUSINESS sounds a bit like a snake eating its own tail. But really, how could it be any other way?

If you're sold on the idea that systemising your business is critical to maximising performance, you intuitively know you must follow a system to systemise your business.

Of course, when I say *you* follow, what I really mean is that your Systems Champion is following the system. Because this is the part where you step aside and allow them to take the reins. As tempting as it may be to micromanage their work or continually look over their shoulder, it's important that you give your Systems Champion the time, resources and authority they need to learn the role, make some mistakes, figure out how to overcome obstacles and eventually make a success of the project. They will be doing the bulk of the work, so you need to give them room to do it.

Which isn't to say your responsibility ends here.

I encourage you to read the remainder of this book before you hire your Systems Champion. Firstly, so that you have a crystal-clear idea of what skills are needed in the role, and secondly, so you know what to expect from their output and have a way of measuring their progress.

Yes, you'll need to give them some time to establish their own system of working, but it's still your duty to keep an eye on what your Systems Champion is accomplishing and to measure the effects.

In practice, as you'll see over the coming chapters, you will still be involved in the project from a top-level perspective. This will mainly take the form of strategy and priority discussions with your Systems Champion, accompanied by regular updates on their progress. But arguably even more important is the way in which you support your champion in their work and demonstrate to all of your team your commitment to systemisation.

A few years ago, I had the privilege of working with Michael E. Gerber (perhaps best known for his book *The E-Myth*, a work which explores the myths and challenges common to small business development) and I sent him an early copy of my previous book, *SYSTEMology*. He gave me some valuable feedback. The most important, in my opinion, was this comment:

"You can't let the business owner off the hook."

You see, I had it in my head that I wanted to get the business owner out of the way as quickly as possible so the Systems Champion could get on with making the systemising happen. And while there's some merit to that – the SYSTEMology process goes faster the more the business owner is able to step back and let their champion take charge of the role – Gerber was saying I shouldn't position this as the business owner effectively washing their hands of the project.

You're not going to be involved in the actual hands-on process of documenting, organising and optimising the systems in your business, but you'll still play an important role in helping your Systems Champion be successful. And one of the key elements is being visibly and audibly behind the work your champion is doing.

I've already hinted at this but, for reasons I'll come to later, not everyone in your business is going to be enthusiastic about the systemisation process, and some may outright oppose it. You need to make it clear, from day one, that you are 100 percent behind this project, that it's important to the long-term health of the business and that you expect everyone to give your champion their support.

There's a balance to be struck here. On the one hand, you want your Systems Champion to have the autonomy to manage the project in their own way, and you want to avoid becoming too involved to the point where it becomes a time sink. On the other hand, you can't abdicate responsibility entirely.

It's also important to note that, as the SYSTEMology project progresses, there may be opportunities for you to become more involved. For example, as systems are documented, you may spot some areas in which optimisation is insufficient, and where a process may need completely re-engineering from scratch. Depending on the value to your business, developing an entirely new process may in fact be a very good use of your time.

But for now, this is your stop.

The next chapter, and all the ones that follow, are a guide for your Systems Champion. This is a step-by-step guide to the practical, hands-on work of making SYSTEMology a reality and it's the champion that is going to be carrying out this crucial work.

The results are going to be extraordinary.

SYSTEMology works 100 percent of the time when it's properly applied by a person who is clear on their vision of improving their business. After only a few months, you'll notice that the positive effects of systemisation begin to compound. Each system that is documented and optimised adds a percentage of improvement, and the resulting benefits stack and accelerate.

Let me leave you with a simple example to show you exactly what I mean. Imagine you run a small service business, which looks like this:

- You have 5000 potential customers per year.
- You convert 25% of those leads (one in four leads becomes a customer).
- Giving you 1250 customers (5000 × 25%).
- Each customer purchases from you two times per year.
- Your average sale is $400 per transaction.
- Giving you an annual revenue of $1,000,000 (1250 × 2 × $400).
- Your profit margin is 25%.
- Giving you an annual profit of $250,000.

Now, here's where the magic of systems becomes clear. Through systemisation (turning your best lead generation practices into repeatable processes, capturing your top salesperson's approach for everyone to follow, creating consistent customer follow-up systems, standardising service delivery, and reducing errors through documented procedures) you get a 10 percent improvement in each area.

This is what it looks like after systemisation:

- You now attract 5500 potential customers per year.
- You convert 27.5% of those leads into customers.
- Giving you 1513 customers (5500 × 27.5%).
- Each customer now purchases from you 2.2 times per year.
- Your average sale has increased to $440 per transaction.
- Giving you an annual revenue of $1,463,212 (1513 × 2.2 × $440).
- Your profit margin has improved to 27.5%.
- Giving you an annual profit of $402,383.

The results are extraordinary:

- Revenue increased by 46% (from $1,000,000 to $1,463,212).

- Profit increased by 61% (from $250,000 to $402,383).

- That's an additional $152,383 in annual profit!

This demonstrates the compounding power of systems. You didn't need to double anything or make dramatic changes. Just very achievable 10 percent improvements across a handful of key areas created a 61 percent increase in profit.

This is why I'm so passionate about this stuff! These numbers aren't theoretical. Through proper systemisation, 10 percent gains in each area are not only achievable, they're conservative. And when you consider adding AI into the mix, you'll make even bigger gains. Expect to see small improvements in the short term, but look out for the eventual point of critical mass when it will feel like the entire business has suddenly jumped into a whole new level.

Case Study	Orchestrated Business Success

Like any masterful performance, building a successful business requires more than raw talent. It demands harmony between passion and process. For Alison Rogers, founder of Vocal Manoeuvres Academy, this lesson would transform not just her business, but her entire approach to leadership.

The business owner's challenge

Alison had been teaching singing since she was 15, building her academy from the ground up with an unwavering commitment to nurturing authentic voices. But like a gifted soloist attempting to conduct an entire orchestra simultaneously, she found herself stretched thin, trying to be everywhere at once.

"I had this constant state of heightened anxiety," Alison recalls, her voice carrying the weight of those challenging times. "Always on tools, always working, burying my head in the sand when tasks felt too enormous." Despite Alison's deep musical expertise, her business was an overwhelming cacophony rather than a harmonious symphony.

Discovering the systems approach

After discovering *SYSTEMology* through an Audible recommendation, Alison recognised the potential for change. She began dedicating every Tuesday morning to working on her business's systems, creating sacred sessions that not even her children dared interrupt. But she was still thinking like a soloist, doing the work herself rather than being the conductor orchestrating and leading things.

The turning point came when she chatted with a SYSTEMologist who insisted she needed a dedicated Systems Champion.

Initially, Alison resisted. *I've got to recruit and I've got to interview*, she remembers thinking. *I don't have time for this. I'm doing SYSTEMology for a couple of hours every week. I'll just make it four or five hours and get it done.*

After a few more weeks of hard slog, Alison finally decided it was best to "trust the process" and use the plug-and-play job ad and position description (we provided) for a Systems Champion. The simplicity was beautiful. "I swiped it completely," Alison admits with a laugh. "I just filled in the blanks with my business name, and it snagged me the most incredible Systems Champion."

The Systems Champion impact

The Systems Champion turned out to be exactly what Alison's business needed. They brought a fresh set of eyes to the business without any judgement about what was or wasn't working, creating a safe space where Alison could openly address the broken parts of her business. Through structured documentation and a systematic approach to problem-solving, they began capturing the essence of Alison's unique teaching methodology.

What had previously lived only in Alison's head was now being carefully extracted and standardised, making it possible for others to deliver the same high-quality experience to students.

"It was like having someone who could read my mind," Alison reflects, "but also organise my thoughts in a way that made them accessible to everyone else."

Transformational results

The transformation was profound. Within six months, Vocal Manoeuvres was performing in virtually every premier venue in Australia, managing ensembles of six to 40 people at a time. More importantly, these high-profile events no longer caused the rest of the business to screech to a halt.

"Now the business just hums along like the most beautiful machine," Alison shares.

The key insight

Like a perfectly executed symphony, successful business systemisation requires every element to work in harmony. This isn't just about creating better systems but about orchestrating a business that performs at its peak without missing a beat, even when you're not centrestage. So, if you're ready to stop being a one-person band and start conducting a symphony of success, it's time to find your Systems Champion.

Watch the full interview here:

www.SystemsChampion.com/resources

Business Owner Action Items

☐ Create your Systems Champion position description. Include key responsibilities and the eight essential qualities required for success.

☐ Write a compelling job advertisement that will attract the right talent, focusing on growth opportunities and core competencies.

☐ Determine your initial budget. Will this be a half-a-day-per-week or a full-time role? Remember, the scope of your investment will directly impact how quickly you'll see results.

☐ Review your current team for internal candidates who demonstrate Systems Champion qualities. Consider apprentices and returning team members.

☐ Schedule initial discussions with your team to introduce the Systems Champion concept and gather input on potential internal candidates.

☐ Define clear boundaries between your role as business owner and your Systems Champion's responsibilities to ensure proper empowerment and support.

☐ Set clear time allocation expectations for yourself. Determine specific hours per week dedicated to systemisation and ensure this commitment is protected from other responsibilities.

☐ Plan your resource investment. Consider what tools, training and support you'll provide to help your Systems Champion succeed (software, courses, mentoring).

☐ Create a realistic budget for implementation. Factor in not just salary but also necessary tools, training and potential external support needed.

And relax, I know what you're thinking: *Great, just what I need ...
more tasks on my to-do list!* The good news? Your soon-to-be Systems
Champion can help with many of these items. Your main job right now
is finding the right person to drive this forward. After all, isn't delegating
tasks like these exactly why you're reading this book?

FOR THE SYSTEMS
CHAMPION

Summary

Congratulations. You've been chosen as your organisation's missing piece. Your eye for detail, knack for organisation and ability to see connections make you the ideal Systems Champion. This isn't just another task – it's a career-defining opportunity!

You will discover building successful systems requires more than just documentation. It demands the right mindset, approach and understanding of human behaviour. The key to transformation lies in balancing consistency with quality while effectively managing team resistance to change.

Highlights covered in these chapters include:

- The business's core mission to transition from individual knowledge dependency to documented, scalable systems.

- The seven stages of SYSTEMology that provide the foundation for business transformation.

- Why consistency trumps perfection in building scalable business systems.

- The McDonald's principle: how predictable processes build customer trust.

- The three core reasons people don't follow processes: knowledge gaps, unclear ownership and resistance.

- The Three Pillars framework: Documentation, Tools and Culture.

- How to transform from individual knowledge dependency to systems-driven operations.

- The critical balance between implementation and team adoption.

5

This Book Is for You

GREETINGS, SYSTEMS CHAMPION. WELCOME TO your new role. If you're reading this, it's because the owner of the business has identified you as someone who has the right skills and temperament for what is going to be the most exciting project you've ever worked on. You're going to help champion a cultural shift toward business systemisation across the entire company.

This is a crucial job for helping the company to grow and prosper, and the business owner is putting a huge amount of faith in you to make a success of this. But remain calm. This book is going to walk you through everything you need to know, and no one is expecting you to make miracles happen in just a few days or weeks. This isn't a project that can be rushed and there's some planning to do before you can really start to affect big change.

But let's not get ahead of ourselves. Let's zoom out and talk about the endgame. What is it that you've been hired, or assigned, to achieve in this role[3]?

---

[3] If you've read the preceding chapters aimed at the business owner, you'll already have a fair idea. If not, don't worry, it isn't necessary. The remainder of the book is going to give you everything you need to succeed.

To answer that question, you need to understand where the business is currently at and what's slowing its progress. In its current state the company relies heavily on key individuals who have the knowledge and skill to manage and take care of specific parts of the business. There are some amazing, talented people keeping the wheels turning, and you're going to get to know all of them over the coming months.

But here's the problem …

Relying on specific people to keep everything running smoothly has serious limitations. What if someone who is critical to the operation of the business gets sick, goes on vacation or even leaves the company? Additionally, because in many cases there is only one key individual (aside from the business owner) who knows how to care for their portion of the work, how do we figure out if there are more efficient ways of doing things?

This is a challenge the business owner has been wrestling with more or less since day one. And, to be fair, this is the challenge that all growing businesses run into. The difference is that most other companies just muddle through and make the best of it, whereas you're going to tackle this obstacle head-on.

Here's your overall goal. Read this over and over until it's burned into your brain and becomes the driving force for everything you do in this project:

> **The business must transition from being a business dependent on individual knowledge to one driven by documented, scalable systems.**

We're going to break this goal down into a series of steps, but it's really important that you understand what the ultimate destination is. Whenever you feel like you're hitting a wall or you're getting bogged down in the details, come back to this primary goal. It will help you find the best way forward.

If we really zoom out and take a 30,000-foot view of the project, your goal can be broken down into two elements:

1. Systemise each part of the business so that clients receive a consistent result that can be repeated endlessly and at scale.

2. Remove key person dependency so the business isn't overly reliant on specific individuals and won't fall into chaos if one of them becomes unavailable.

Think of it this way: you're creating a comprehensive playbook for how your business operates, mapping out everything from attracting customers to delivering exceptional service and covering all the crucial steps in between. This playbook, made up of a collection of systems and processes, will allow the business to achieve the following:

Consistency and quality: Systems ensure work is consistent, so that customers get the same quality of product or service every single time, no matter who's handling the tasks. Showing customers that you can deliver – to a high standard, every time – is how you build trust and loyalty in the marketplace.

An empowered team: With well-defined processes, team members know exactly what to do and what results are expected of them, which can lead to reduced training and supervision requirements. This gives them confidence to take ownership of what they do and deliver their best work.

Efficiency and reduced errors: No business will ever be perfect, but when every process is proven and followed closely, mistakes are minimised and efforts are not wasted. Fewer errors and better efficiency ultimately save time, effort and resources.

Easier and more enjoyable work: Well-designed systems make work easier, more pleasant and less frustrating. They help prevent the business from stalling when a key team member is unavailable. No more late nights and weekends putting out fires because someone gets sick or leaves unexpectedly!

Scalability: The business has big ambitions to grow, but it will always be limited if it has to constantly reinvent the wheel or rely on a few key individuals. Once it has a clear roadmap that others can follow, it will be able to expand smoothly and efficiently.

You're going to work closely with the teams within the business, across all departments, capturing the knowledge that's currently in their heads and translating it into documented systems that anyone, to a reasonable degree, can follow. Not necessarily with the same smoothness and expertise as the person that does the job day in and day out, but well enough to keep the machine running.

Again, you can take a deep breath if this feels like an impossibly ambitious goal. Your goal isn't to turn your business into a perfect machine. All you're required to do is put the pieces in place so that, at the end of the project, you have a solid "Version 1.0" of a systemised business. It will be far from perfect, and that is to be expected. This version is then going to be revised and improved, repeatedly, over a period of years, each time getting a little smoother, a little more efficient. You may or may not be involved in this iteration process; your job is simply to create the first pass.

And you won't be doing this alone. The business owner is going to be backing you all the way and will assist you with some of the more difficult decisions. Or, at least, they should.

Some business owners who employ a Systems Champion fall into one of two extremes. They either try to micromanage and look over your shoulder every step of the way, or they drop everything into your lap and forget about it. Most will happily sit where they should – in the middle ground. They'll trust you to get the job done but take an active interest in your progress and be ready to help out when challenges arise.

If they're a micromanager, you need to ask for the resources and authority to get on with the job without having to "bother" them all the time. Remind them that one of the subgoals of this project is to help them step back from the day-to-day running of the business, and that this will never happen if they don't authorise you to push forward.

For the opposite, an owner who doesn't make time to support you, remind them of the same and that it's important they have some input and keep an eye on your progress. Push for a daily meeting (these can be short, maybe 10–15 minutes) over the first couple of weeks, moving to weekly or bi-weekly meetings thereafter. Use these meetings to update the business owner on your progress and to solicit opinions on key decisions.

This isn't a straightforward project. You've been chosen because you've demonstrated that you're a resilient problem-solver who enjoys a challenge. But don't make the mistake of thinking that the business owner is expecting you to create something with the precision of a Swiss watch. For now, a Casio or a Timex is absolutely fine.

You're going to find this to be challenging but incredibly rewarding work that will have a lasting impact on how the business operates. You're going to play a key role in driving the company to the next level of success!

You probably have a million questions, and that's good. But don't worry about the details for now. My recommendation is to start by reading this book in its entirety, making as many notes as you can, so you have a clear overview of how this is going to work in practice. Once you've done that, you can return to the start and begin working through the steps, one at a time.

Just remember to enjoy the process! You're playing a crucial role in shaping the future success of the business. Your work is incredibly important, not just to the business owner, but to everyone at the company who is depending on the business to keep flourishing. The satisfaction you're going to experience over the coming weeks and months will be unlike anything you've previously experienced, and this is your opportunity to show everyone what you're capable of.

You're going to do great!

Make the Goal Your Own

Your goal is: "The business must transition from being a business dependent on individual knowledge to one driven by documented, scalable systems."

How would you explain the goal in your own words?

....

....

....

Why do you think this goal matters for:

The business: ....

....

....

The team: ....

....

....

The customers: ....

....

....

Please note: This book was designed to be a workbook. Write in it, highlight it, add sticky notes … make it yours! The more you engage with these pages, the better your results will be.

6

SYSTEMology Explained

BEFORE WE GET STUCK INTO the meat and potatoes, I think it will be helpful for you to understand a little more about the foundations from which this book was written. *Systems Champion* builds upon a previous book I've written called *SYSTEMology*.

Now I know what you're thinking, and the short answer is: no, you don't need to read that book. The book you're reading now stands on its own, but let me give you a quick crash course anyway.

SYSTEMology details my seven-stage system for systemising a business. At a high level the stages are:

Stage 1: Define – Select the 10–15 systems to document first. This stage aims to answer the crucial question of where to start systemising your business.

Stage 2: Assign – Identify who in your team already knows how to complete these tasks to a great standard.

Stage 3: Extract – Quickly and easily get the knowledge out of team members' heads and into documented form.

Stage 4: Organise – Store this information centrally so that the entire team can access it.

Stage 5: Integrate – Get your team on board and excited by the idea of systemisation.

Stage 6: Scale – Identify what other systems are required to scale the business beyond what was identified in Stage 1.

Stage 7: Optimise – Re-engineer your processes to get more done with less, continuously improving your business.

Of course, the book goes into much more detail than this, but you get the gist. When you hear me mention "rolling out SYSTEMology", I simply mean applying these ideas for systemising to your business. This includes all my tools, tactics and strategies. Just be aware, while I've presented the seven stages in a logical sequence here, in reality that implementation isn't always linear. Every business is different, with its own unique challenges, culture and starting point.

As a Systems Champion, you'll learn to adapt the methodology to fit your specific situation. Sometimes you'll work on multiple stages simultaneously. Sometimes you'll need to circle back to earlier stages as you discover gaps. That's perfectly normal and expected.

Your job as the Systems Champion is to make informed decisions about where to focus your efforts for maximum impact. I want you to have a much more active role and this is primarily why I changed the way I have delivered the ideas in this book. Everything still fits back into the original seven stages of SYSTEMology – it's just displayed differently.

This means if you've already read *SYSTEMology*, you can always return to it and it will help you find where to focus next. And if you haven't, that's okay too! Toward the end of this book I will help you create your own customised action plan. That said, as long as you remember your core job is to build consistency and reliability through documentation … you'll always find your way.

Consistency over perfection

Let me share a secret that took me years to fully understand. In business, consistency is far more important than perfection. This might sound counterintuitive. After all, don't we all want to deliver the absolute best? But here's the truth: your clients value reliability over occasional brilliance.

McDonald's is the perfect example of this. No one claims they make the world's greatest hamburger. They probably wouldn't even make your top 10 list of favourite burgers. Yet they've built one of the most successful small businesses in history. Why? Because whether you're in Sydney, Tokyo or New York, you know exactly what you're going to get.

This predictability is incredibly powerful. When a customer walks into McDonald's, they're not hoping for a culinary masterpiece. They're buying certainty. They know the Big Mac will taste the same as it did last time. The fries will be exactly how they remember them. The service will follow a familiar pattern.

Now, I'm not suggesting you turn your business into McDonald's. In fact, I believe blindly following McDonald's' way of doing business can do more harm than good. But there's a profound lesson here about what clients truly value: they want consistency.

Consider this. Would your clients prefer a service that's brilliant 20 percent of the time and merely okay 80 percent of the time, or one that's consistently good 100 percent of the time? In almost every case, they'll choose consistency. Why? Because consistency builds trust, and trust is the foundation of every successful business relationship.

This is where systems come in. A well-designed system is simply a series of steps that, when followed, deliver a consistent outcome. It's not about achieving perfection but about reliability. Every time someone in your business follows a system, they should achieve a similar result. This repeatability is what allows you to scale, to train new team members and to deliver reliable results even when you're not personally involved.

And this is precisely where SYSTEMology shines. By following the seven-stage framework I outlined earlier, you're not just creating random systems; you're approaching systemisation itself systematically. There's beautiful consistency in using a proven methodology to build your systems. Your team learns that there's a specific way to document, store and improve processes. This meta-consistency (being consistent about how you create consistency) compounds the benefits.

McDonald's took this to the extreme, documenting every minute detail of their operation. But for most businesses, that level of documentation isn't necessary or practical. This is why SYSTEMology focuses on the critical few rather than the trivial many. Consistency should come first, and complexity can follow later when you're ready.

Measuring success

How will you know if your SYSTEMology implementation is successful? There are several key indicators to watch for.

Business owner freedom is the most obvious sign, meaning the owner can step away from day-to-day operations without everything falling apart. This pairs with reduced key person dependency, as no single individual becomes essential for the business to function.

Team confidence emerges as members know exactly what's expected of them, resulting in a consistent client experience where customers receive the same high-quality service regardless of which team member they interact with.

Improved metrics provide quantifiable evidence of success as systems reduce errors and increase efficiency. A cultural shift occurs when "this is how we do things here" becomes a common saying throughout the organisation.

Clearly the rewards are worth it. That's why the business owner wants to make this a priority. Now, I will do my best to give you everything you need in this book, but should you wish to go deeper, you may also wish to check out the original *SYSTEMology* book. I will leave this to you to decide.

7

Overcoming Resistance

YOUR GOAL AS SYSTEMS CHAMPION is to build a fully systemised business. But between you and that goal stands one significant challenge … and it's not documenting the systems themselves.
It's bringing your team along with you.

Even though you're excited to improve things for everyone, the reality is that change can be scary. Your teammates are used to doing things
their way, and not everyone will jump at the chance to document their
processes or adapt to new methods.

Over the years, as you could imagine, I've heard hundreds of excuses
as to why teammates don't follow processes. But when closely examined,
these objections boil down into just three core excuses. They are the
fundamental barriers that stand between Systems Champions and their
vision of a well-ordered business.

The reality is, you can develop the most comprehensive, perfectly
documented systems imaginable, but if your team doesn't use them,
they're merely digital paperweights. Your success hinges entirely on
team adoption. Let's address these core excuses head-on:

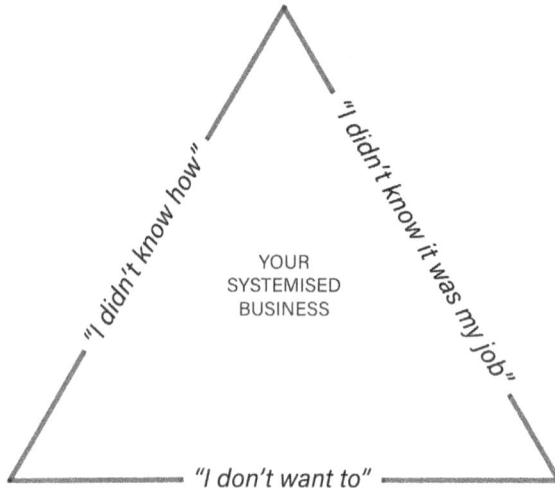

Excuse #1: "I didn't know how"

This is the most common excuse, and it's exactly what it sounds like. Team members simply don't know there's a process to follow. We see this manifest in several ways. Proper training might be missing. Documentation could be unclear or hard to understand. Team members might not have access to the right information when they need it. Sometimes it's as simple as confusion about the specific steps involved, or they might not even know what "good" looks like. What does success mean for this task?

This is actually the easiest excuse to address because it's purely about information and access. Think prevention, not reaction. The goal is to make clarity the default state of your business. Put systems exactly where people need them, when they need them. Make accessing information easier than asking for help and ensure it's never more than one click away at the moment it's needed.

Having well-organised systems management software becomes critical here. This creates that single source of truth where team members can instantly find the guidance they need.

Excuse #2: "I didn't know it was my job"

This excuse goes beyond just knowing how to do something. It's about clarity of responsibility, and it's more common than you might think. You'll notice it when tasks aren't being completed on time or when work starts falling through the cracks.

There's often a general lack of accountability, and you'll frequently hear confusion about who owns what parts of a process. This challenge typically arises due to unclear roles and responsibilities, and while it takes more effort to address than the first, it's still relatively straightforward to solve by installing the right tools.

Leverage project management software. Create unmistakable hand-off points between team members. Build checkpoints into your processes that make expectations impossible to miss.

Excuse #3: "I don't want to"

This is the trickiest type of resistance to tackle because it's really about motivation and engagement. You'll hear various versions of this excuse: "I don't have time," "I don't have the resources," or "The old way works fine." But here's the truth: these are rarely about actual time or resource constraints. They're about priorities.

When someone is truly motivated to do something, they find a way to prioritise it. This is the most challenging resistance to overcome because it requires addressing mindset and engagement, not just processes and procedures.

Now, with each of these three excuses there can be many other variations. However, when you boil things down, you will find 99 percent of all excuses will fall into one of these categories.

It's important to understand at this point that as a Systems Champion, you're not the company's disciplinarian. Your role isn't about enforcing rules or catching people doing things incorrectly. Instead, you're there to create elegant systems that naturally bring clarity and structure to the workplace.

When done right, these systems make non-compliance obvious without you having to play detective. They give managers the practical tools they need to address issues efficiently. Most importantly, they set every team member up for success from day one. Think of yourself more as an architect of success than an enforcer of rules. You're creating an environment where good processes naturally lead to good outcomes.

Resistance Reality Check

Working with the business owner, note where you've already noticed each type of resistance in your business:

"I didn't know how"

Examples I've seen: ....

....

....

Potential solutions: ....

....

....

"I didn't know it was my job"

Examples I've seen: ....

....

....

Potential solutions: ....

....

....

"I don't want to"

Examples I've seen: ....

....

....

Potential solutions: ....

....

....

Which type of resistance do you expect to encounter most?

....

....

....

....

8

Three Pillars

I N T H E F O L L O W I N G C H A P T E R S W E ' L L further explore these excuses and how to prevent them using three interconnected pillars. Think of these pillars as the legs of a stool: each one is essential, and they work together to create stability. Master these and you'll dramatically reduce resistance while creating a foundation for lasting change.

They hold the key to building your systemised business.

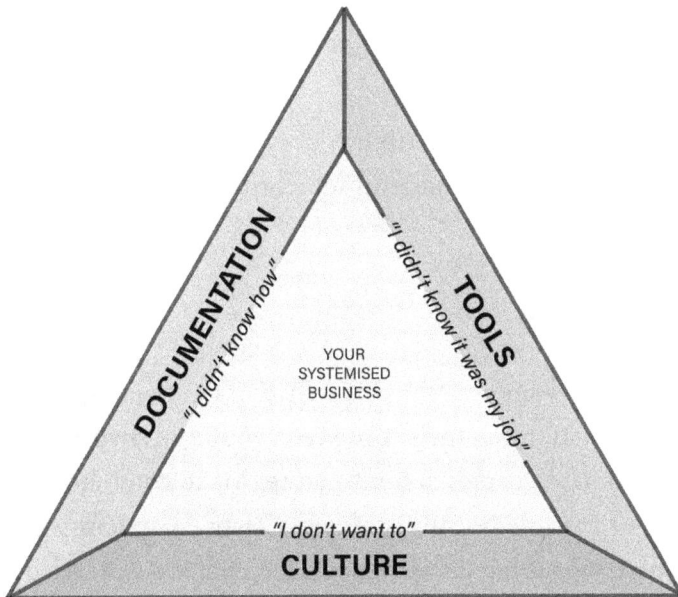

Pillar 1: Documentation

Documentation is your foundation and abolishes the "I didn't know how" excuse. It transforms tribal knowledge into clear, actionable guidance: crystal-clear systems that leave no room for confusion. When your documentation is accessible and well-structured, managers can confidently point to any process and say, "This is how we do it here."

The goal isn't just to document. It's to create instructions clear enough that anyone with a basic understanding of your business can follow them and achieve consistent results. If a team member with relevant skills is struggling to understand a process, that's usually a sign your system needs refinement. Remember, we're aiming for clarity that enables competent team members to succeed, not trying to oversimplify complex work.

Pillar 2: Tools

Tools create the infrastructure for success and remove the "I didn't know it was my job" excuse. It's not just about software but about creating accountable transparency. Your tech stack should make it crystal clear who's doing what by when. It should make responsibilities visible to all and track progress in real time.

The goal is to ensure the team is always able to seek out the most efficient ways to get tasks done and autocorrect when team members fall off track.

Pillar 3: Culture

Culture is where your systems come alive. It's not enough to just have great documentation and tools. You need to build an environment where systems thinking becomes second nature. Make continuous improvement part of your daily conversations. Celebrate system wins. Create spaces where suggesting improvements feels natural and welcome.

Getting this third pillar right will help to dissolve the "I don't want to" excuse. Once systems become embedded in your company's DNA, resistance fades and your new culture begins to take hold.

The truth is, people will do far more to fit in with their peers than for any other motivator. By transforming "how we do things here", you'll make it easier for others to join in. Changing existing culture can be challenging at first – you're asking people to shift how they view their work and the business. But this gets easier over time, especially as new team members join who are naturally aligned with your systems approach.

The power of integration

Here's one important thing to understand: these pillars don't work in isolation. They're deeply interconnected, each supporting and enhancing the others.

The best documentation in the world is useless if your tools make it hard to access. The finest tools become shelfware if your culture doesn't embrace using them. And trying to build a systems culture without proper documentation and tools is a bit like trying to run a restaurant with no recipes or kitchen equipment.

In the following chapters, I'll provide you with a range of powerful tactics and strategies to strengthen each of these pillars. In truth, there are many different approaches you could take, but I've applied the 80/20 rule to give you only the most powerful techniques I've discovered that will deliver the greatest results. You don't need to implement everything at once. Focus on reinforcing the weakest pillars first, then build from there.

Later in this book, you'll develop a detailed action plan with clear, linear steps. But where you begin with these three pillars (and the precise order you tackle them) will depend on your specific situation. Your approach should be strategic – sometimes you'll need to strengthen your documentation first, other times your tools need immediate attention and in many cases your culture may require the primary focus.

The beauty of this framework is its flexibility and focus. Rather than trying to systemise everything at once, you'll develop a plan that targets your specific needs. When you approach implementation this way, your business transformation becomes not just possible but inevitable.

Pillar Strength Assessment

Rate your current foundation in each area:

Pillar 1: Documentation

Weak → Developing → Strong

Rating 1–10: ....

What's missing? ....

....

....

Pillar 2: Tools

Weak → Developing → Strong

Rating 1–10: ....

What's missing? ....

....

....

Pillar 3: Culture

Weak → Developing → Strong

Rating 1–10: ....

What's missing? ....

....

....

Which pillar needs immediate attention? Why?

....

....

Case Study — Finding a Systems Hero

When Renee Kelly started Lime Therapy 15 years ago, she was a solo occupational therapist working from her farm in Mildura. Today, her allied health practice employs 40 team members and serves thousands of clients. But like many successful businesses, their growth created challenges when it came to maintaining quality, ensuring consistency and making everything sustainable.

"I'm a big-picture person," Renee admits. "I'd get excited about ideas and say, 'Let's do this!' and someone would ask, 'But how?' and I'd say, 'Don't worry about that!'" And while this approach had taken them far, they knew something was missing.

The turning point

The breakthrough came unexpectedly. During an online program she was attending, Renee heard a young therapist mention SYSTEMology. "A light bulb went off," she recalls. "If only I'd known about this when I was starting out!" She immediately ordered the book – one copy for herself and one for her husband, Matt.

But here's where Renee and Matt made a crucial decision that would transform their business. Instead of trying to implement systems themselves, they looked for someone who could champion this change within their organisation. They found their answer in Kaleb, a young occupational therapist who'd been with them for just two years.

The perfect Systems Champion

What made Kaleb ideal wasn't years of experience or technical expertise. It was his natural inclination toward organisation and his genuine curiosity about how things worked. "I like things organised," Kaleb explains. "I like my lists, and I like to know what to do."

Renee and Matt gave Kaleb something precious: time and space to focus on systems. They protected his hours for this work, ensuring he could meet with team members and document processes without trying to squeeze it in between client appointments.

The impact

The results were transformative. Kaleb approached his role with enthusiasm, working closely with experienced team members to document their knowledge. He created a system for sharing updates through team chat, keeping everyone informed and involved. Most importantly, he made systems exciting.

"People started coming to me saying, 'I want to do what Kaleb's doing. I don't know what he's doing, but it looks fun!'" Renee shares. The team began viewing challenges through a systems lens, actively seeking ways to improve their processes.

The business saw immediate practical benefits, like reducing their invoicing time tenfold. But the real transformation was cultural. Systems thinking became part of their DNA. As Renee notes, "SYSTEMology has become part of our language. Every problem, every opportunity, we now see it as a system."

Lime Therapy's story reveals a powerful truth about the Systems Champion role. It's not about finding the most experienced or highest-ranking team member. It's about finding someone with natural organisational tendencies, genuine curiosity about how things work, strong communication skills and fresh eyes to see opportunities others might miss. This combination of qualities, often found in up-and-coming team members like Kaleb, can transform your systems initiative from a management directive into an organic cultural movement that energises your entire team.

For Kaleb, this role offered an incredible opportunity to learn the business from the inside out while making a significant impact. For Renee and Matt, finding the right Systems Champion turned their systems dream into reality.

Watch the full interview here:
www.SystemsChampion.com/resources

Systems Champion Action Items

☐ Schedule a meeting with the business owner to understand their vision for consistency across the business. What level of standardisation are they hoping to achieve?

☐ Create a dedicated space (digital or physical) for organising your Systems Champion work. You may wish to include a journal to track your progress and insights.

☐ Define your interpretation of the core mission statement: "The business must transition from being a business dependent on individual knowledge to one driven by documented, scalable systems." What does success look like in your specific organisation?

☐ Set up initial conversations with key team members to understand their current ways of working. Note which processes show the most variation in outcomes. Focus on listening and learning, not changing things yet.

☐ Start documenting examples of the three types of resistance ("I didn't know how," "I didn't know it was my job," "I don't want to") as you observe them in the business.

☐ Review your current tools and systems. What documentation already exists? Where is it stored? How do people access it?

☐ Begin drafting your 90-day plan for strengthening the three pillars in your business – Documentation, Tools and Culture.

☐ Optional: Add the original *SYSTEMology* book to your reading list to review once you're done with this one.

Remember: you don't need to have all the answers yet. Your job is to start building a foundation for success. The journey of a thousand miles begins with a single documented process!

PILLAR 1: DOCUMENTATION

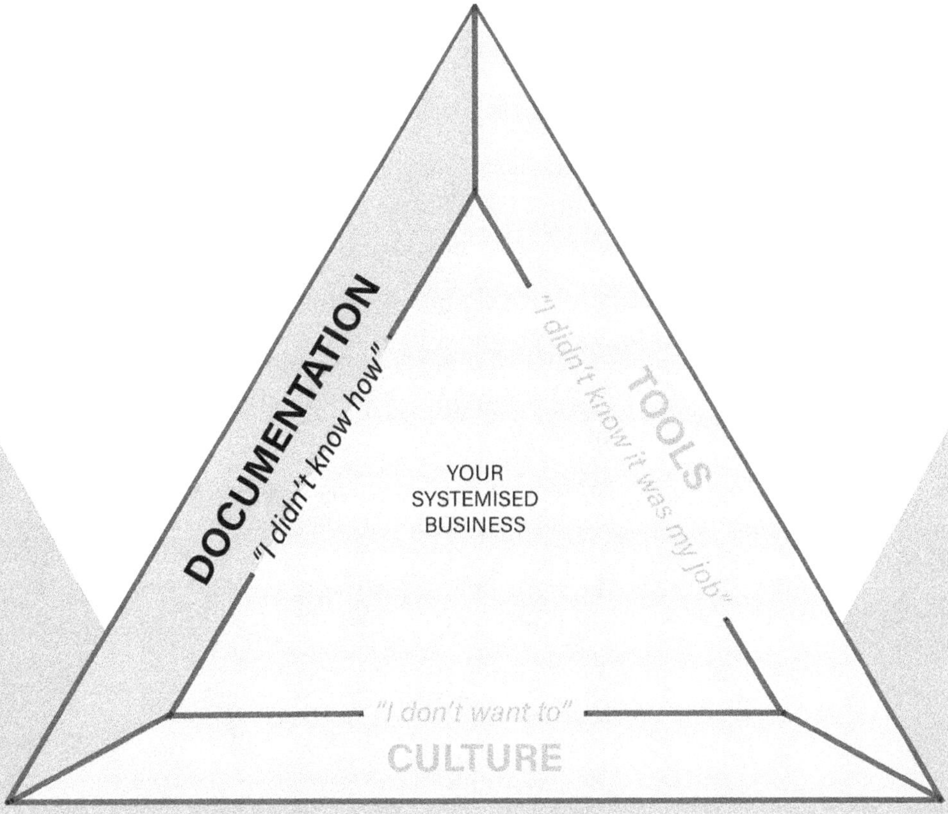

DOCUMENTATION
"I didn't know how"
TOOLS
"I didn't know it was my job"
YOUR SYSTEMISED BUSINESS
"I don't want to"
CULTURE

"Without standards, there can be no improvement."

Taiichi Ohno, father of the Toyota Production System

Summary

Successful systems documentation requires more than just writing down steps. It demands a strategic approach that balances consistency with usability. The key is treating system creation as a craft, combining both art and science to capture what makes your business work.

Highlights covered in these chapters include:

- The McDonald's principle: how systematic documentation transformed a small restaurant into a global empire.

- Why documented habits are the foundation of business scalability.

- The updated System for Creating Systems 2.0 framework that leverages AI for efficient documentation.

- Eight practical rules for effective system extraction and documentation.

- The critical balance between simplicity and completeness in system design.

- How to transform tribal knowledge into repeatable processes that anyone can follow.

- The importance of making systems human-centric rather than purely technical.

9

Documented Habits

THINK ABOUT THE LAST TIME you had a Big Mac. Whether you ordered it in New York, Tokyo or Melbourne, Australia, it probably tasted exactly the same. That's no accident. But this story of systematic perfection didn't start in a boardroom. It started with a curious milkshake mixer salesman named Ray Kroc.

In 1954, Kroc was 52 years old and making a modest living selling milkshake mixers when he noticed something odd. A small restaurant in San Bernardino, California had ordered an unusually large number of his machines. Intrigued, he decided to visit the place personally.

What he discovered that day would change business history forever. The McDonald brothers, Dick and Mac, had created something revolutionary: what they called the "Speedee Service System". While other drive-ins of the 1950s were chaotic affairs with carhops, lengthy menus and 30-minute wait times, the McDonald brothers had engineered an operation that worked like clockwork.

Their kitchen was a marvel of efficiency, laid out like an assembly line. Every movement was choreographed, every process standardised. They had stripped their menu down to just a few items, designed custom equipment for consistent cooking and created precise protocols for food preparation. They specified exactly one squirt of ketchup and one of mustard on every burger. The result? They could deliver a consistently perfect meal in under a minute.

But what really caught Kroc's attention was that he could see how the system could be replicated anywhere. While the McDonald brothers were content with their local success, Kroc envisioned bottling their efficiency and reproducing it across the country. It wasn't just about selling hamburgers. It was about selling a proven system of selling hamburgers.

This is systemisation at its finest. They took all the knowledge that traditionally lived in a chef's head (the timing, the temperature, the techniques) and transformed it into step-by-step instructions that anyone could follow. No more "you just know when it feels right" or "it comes with experience". They made the invisible visible.

The challenge most small businesses face is quite scary. Their most valuable asset, their intellectual property, isn't truly theirs – it's trapped in the minds of their employees. Think about it. Every business spends years developing their "special sauce" through countless hours of trial and error. Star performers have honed remarkable skills and insights, developing techniques that reliably produce exceptional outcomes for clients. Yet without proper documentation, this hard-earned expertise (the very DNA of your business's success) remains trapped inside their heads. If they go, so too does that expertise.

So much of your team's daily work, far more than most realise, represents years of accumulated wisdom and experience. These are the habits and practices that have been refined through thousands of hours of real-world testing. Some are brilliantly efficient, some need improvement, but they all share one critical trait: they exist only in the minds of your team members. This means your business's most valuable asset, its operational intelligence, walks out the door every evening.

Documentation solves this. It's the foundation of systemisation, the first pillar that makes everything else possible. Think of it as capturing your company's operational DNA. It transforms all those invisible positive habits into a visible, living playbook that can be shared, refined and scaled.

When you document these habits, something magical happens. You turn implicit knowledge into explicit instructions – a "how-to" that anyone can follow to complete the same work. Will they immediately match the expertise of your seasoned pros? Probably not. But they'll be able to achieve a successful outcome by following the documented process.

Now, am I suggesting you systemise like McDonald's? Not exactly. You need to be careful blindly copying other businesses' strategies because I'm going to guess you're probably not running a hamburger shop. You're probably not employing 15-year-olds who've never had a job before, and you may (or may not) be looking to become a global enterprise.

You're just starting on your systemisation journey and your business probably deals with different levels of complexities and team member skill levels. This doesn't mean documentation isn't important. It just means you need a different approach. You need documentation that matches your business's sophistication while still providing the clarity and consistency your clients crave.

Training new skills and habits

The Conscious Competence Ladder (developed by Noel Burch) describes the stages individuals progress through as they acquire new skills and habits. And it quickly becomes evident when looking at it through the lens of systemisation how documented processes can accelerate the way in which people master new skills.

1. Unconscious Incompetence – "You don't know what you don't know"

This is where everyone starts. New team members don't know what they don't know and that's to be expected. You might even have existing team members who are blissfully unaware that there are better ways of completing their tasks. Sadly, this blindness creates waste and leads to more mistakes than you realise.

2. Conscious Incompetence – "You know what you don't know"

Documented processes that capture best practice reveal to the team members a better way of doing things. Team members become aware that they are not doing things in the most efficient manner. For some, this can create feelings of overwhelm and frustration, while for others, it motivates them to improve. Either way, the first step toward improvement is awareness.

3. Conscious Competence – "You know and do but you have to think about it"

Systemisation done well helps team members dramatically improve their output by consciously following established processes. They gain new skills but it takes effort to keep using them and not slip back into old habits. The individual may need to keep referring back to the system, and use guides, checklists and tools to ensure they're following the process.

4. Unconscious Competence – "You know and do without thinking"

In the final stage, working the systems becomes second nature and the person uses their new abilities effortlessly and without conscious thought. Everything is intuitive and automatic. Team members at this level are the most helpful for improving existing systems and developing new ones.

Understanding where each team member is on this ladder helps you provide the right support in their systemisation journey. Some people you work with will be at the bottom of the ladder and need a gentle introduction to new ways of doing things. Those in stage two will need reassurance and support. At stages three and four, you have team members who are going to be powerful allies in shaping the company's systemisation culture.

Meet people where they are

Start by acknowledging that documentation isn't about criticism but about improvement. Every process you document is an opportunity to make your business better. When team members understand this, they're more likely to embrace the documentation process rather than resist it.

And remember, the goal isn't perfection. It's progress. That's where the real magic of systems lives.

Each system you create adds a small percentage of improvement to your business. A documented client onboarding process might save 10 minutes. A standardised email response system could save five. A simple invoicing process might save 10 minutes. Individually, these improvements might seem modest, but these gains stack and accelerate. Ten systems, each saving just five minutes per use and reducing error rates, could give you back 10 or more working days annually.

But the true transformation happens, when your systems work together and amplify. Just like good habits compound into great results, systems multiply each other's effectiveness. This isn't simple addition – it's multiplication! That client onboarding system is now fuelling your welcome process, tightening your feedback loops and sharpening your service all at once. Each system reinforces and enhances the others, creating a web of efficiency that becomes a force multiplier for exponential growth.

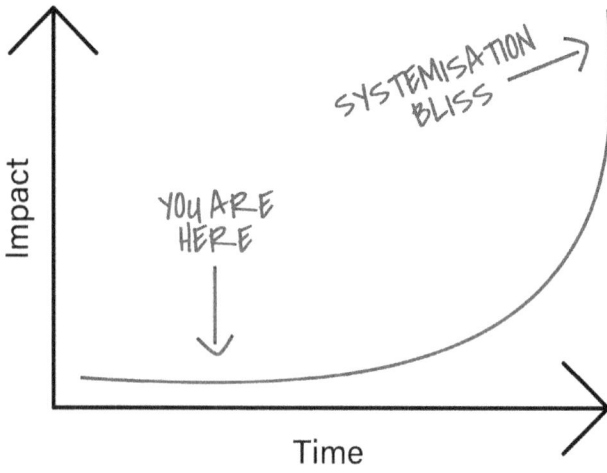

So where do you start with documentation?

10.

Minimum Viable Systems (MVS)

THINK ABOUT THE MIRACLE THAT is the human body for a moment. Right now, as you're reading these words, an incredibly sophisticated set of systems is working in perfect harmony to keep you alive. Your heart beats about 100,000 times every day, pumping blood through roughly 97,000 kilometres of blood vessels. Your lungs will breathe in and out around 22,000 times today, delivering vital oxygen to every cell in your body. Your digestive system will break down food into the exact nutrients needed to keep everything running.

But what's truly extraordinary isn't just these individual systems – it's how they work together. Each system depends on the others to function properly. Your respiratory system needs your circulatory system to distribute the oxygen it takes in. Your digestive system relies on your endocrine system to regulate metabolism. Your muscular system requires your skeletal system for support and movement.

It's obvious when I say it but, if any one of these vital systems fails, the consequences can be severe. A problem with your immune system leaves your whole body vulnerable to attack. Issues with your circulatory system can affect everything from your energy levels to your brain function. In the worst cases, the failure of a critical system can be fatal.

Even less severe issues can have far-reaching effects. A minor problem with your digestive system might affect your energy levels, which impacts your ability to exercise, which then affects your cardiovascular health, and so on. It's all connected.

Your business: a living system

This is exactly why I love using the human body as an analogy for business. Because your business operates in exactly the same way. Just like your body, your business is a collection of vital systems (marketing, sales, operations, finance, HR, management, etc.). They're all interconnected. Remove one, or even just allow it to become inefficient, and the whole business will eventually suffer.

For example, a weak finance department can lead to cash flow issues, which impedes growth, even if every other part of the business is excelling. Conversely, when every department is functioning optimally, the business remains healthy and is capable of thriving, even when external forces put pressure upon it.

The good news is, your business isn't as complex as the human body. Which is just as well since trying to systemise your business to the same level of minutiae as the human body would be an almost impossible task. I've seen what happens when a business tries to systemise everything and it's always a disaster. Aside from it being a herculean task that eventually collapses under the weight of its own ambition, it's impossible to keep up with the ongoing work of updating and improving the systems you do manage to put in place.

So where do you start? Which systems should you document first? This is one of those times it pays to be selective. Heard of the 80/20 rule? It suggests that 20 percent of your efforts produce 80 percent of your results. Well, when it comes to business systems, this principle couldn't be more applicable.

It means you can achieve extraordinary results by focusing your attention on just that critical 20 percent. We call this your Minimum Viable Systems (MVS): the 20 percent of systems that deliver 80 percent of your results. Think of it as identifying the critical tasks your vital organs must accomplish to remain healthy.

There's also another important point to understand here. Just because something isn't documented doesn't mean it magically stops happening. Your business is already functioning without documentation and will continue to do so. Finding your MVS is about being strategic, identifying what's truly critical rather than trying to capture everything.

Is it essential, repeatable and delegable?

There's a knack to identifying the systems that will most benefit the business by being documented. This will come with time and practice. But the above question – "is it essential, repeatable and delegable?" – is a good mantra to keep front of mind.

The first two concepts, "essential" and "repeatable", are pretty self-explanatory. If a system is crucial to the core function of the business and is performed regularly with minimal variation, it's a prime candidate for the standardisation benefits that come from systemising it.

Delegation, however, is a little more complex and is worthy of explanation because using this to help you select which systems to document is going to offer some unique advantages.

One of the long-term goals of SYSTEMology is to systemise processes to the extent that it becomes possible for lower-skilled team members (or machines) to complete them. The result is that more skilled individuals can eventually let go of those tasks altogether and work on more strategic elements of the business, such as problem-solving and driving business growth.

Systemisation can, in effect, move processes down through the chain of experience and ability so they're managed on a lower level. This is great for less experienced workers who can become involved with more interesting and valuable areas of the business. And it's even better for more senior workers who are liberated from repetitive tasks that keep them from spending time on higher value tasks.

Wherever possible, find tasks that fit into all three categories: essential, repeatable and delegable. Prioritise those tasks that are already being successfully completed in the business and avoid activities that are not yet being done.

This exercise is aimed at finding already successful habits and making those repeatable. I think it's easiest if we jump straight in and make this practical. Head to **www.SystemsChampion.com/resources** to download the MVS template.

How to find your Minimum Viable Systems

Step 1: Identify your business departments

Every business has distinct functional areas. These are like the vital organs in your body. Each plays a crucial role in keeping your business alive and healthy. Start by considering these six core departments that exist in virtually every business.

Marketing: This is your business's voice to the world, responsible for getting your message out to potential clients and generating leads. Your marketing department handles everything from creating content and advertising campaigns to broadcasting emails to your database, all designed to attract your ideal clients.

Sales: Once marketing has attracted potential clients, your sales department takes over. They're responsible for converting interested prospects into paying clients. This includes managing your entire sales process, from creating proposals and following up with leads to bringing new clients smoothly into your business.

Operations: Think of operations as the engine room of your business. This is where your products or services get delivered. Your operations team ensures quality control, manages client relationships and oversees all the processes that deliver value to your clients.

Finance: This department manages the flow of money through your business. They handle everything from sending invoices and collecting payments to managing expenses and monitoring cash flow.

Human Resources: This department ensures you have the right team in place. They oversee recruitment, onboarding and parts of performance management, while also nurturing your team culture and maintaining clear internal communications.

Management: Like your brain coordinating your body's activities, your management department provides direction and oversight for the entire business. They handle strategic planning, business development and performance monitoring, ensuring all departments work together effectively toward common goals.

> **PRO TIP:** Don't worry if some functions overlap or if one person handles multiple departments. This is normal, especially in growing businesses. The goal is to identify functions, not create rigid divisions. Every business has these core functions, even if they're not formally structured this way yet. Understanding these departments exist gives you the framework you need to build your Minimum Viable Systems.

Step 2: Identify your core systems in each department

In this step you're going to systematically work your way through each of the identified departments with a specific line of thinking: "If I had to pick the most critical seven tasks going on in this department, that are essential to this department being a success, what would they be?"

It is recommended you do this in conjunction with either the business owner and/or department head if/where this person has been nominated. These team members will help you prioritise based on their knowledge of their respective departments.

If you get stuck it might be helpful to consider under what cadence these tasks occur. Ask what tasks may happen on a triggered basis (e.g. taking a phone call) as opposed to others that may occur at specific intervals like daily, weekly, monthly, quarterly or annually (e.g. creating a monthly report).

Work your way through each department and each cadence, asking: "If I had to pick seven systems for this department, what would I pick?"

Department: ....

Triggered tasks:

....

....

....

Daily tasks:

....

....

....

Weekly tasks:

....

....

....

Monthly tasks:

....

....

....

Quarterly tasks:

....

....

....

Annual tasks:

....

....

....

Obviously there are many more tasks that occur, but for now, try to be as particular as you can. The challenge here is always prioritisation. Your final deliverable should be a focused list of no more than seven systems per business department, creating a manageable total of 42 systems across your entire company. On the next couple of pages, you will find both some completed MVS templates and an empty MVS template you can use to draft yours.

Example: Residential Home Builder

Minimum Viable Systems™ (MVS)

1 Marketing	2 Sales	3 Finance	4 Human Resources
Lead Generation via Website & Local SEO	Initial Client Consultation & Qualification Script	Client Progress Claim & Invoicing Process	Recruitment & Interview Process
Google Ads Campaign Setup & Management	Estimating & Quoting Process	Subcontractor Payment Workflow	Site Staff & Office Onboarding Workflow
Lead Capture & CRM Entry Process	Proposal & Contract Creation	Supplier Invoice Entry & Approval	Employee Licensing & Compliance Check
Project Updates (Portfolio Management)	Sales Pipeline Management (CRM Updates)	Weekly Cash Flow Forecast Review	Weekly Toolbox Talk Process
Review & Testimonial Collection Process	Pre-Start Meeting with Clients	Payroll & Timesheet Processing	Incident & Safety Reporting Workflow

Referral Program Workflow

Monthly Marketing Performance Review

Job Handover from Sales to Operations

Sales Reporting & KPI Review

Job Costing & Variance Reporting

BAS & Compliance Submission Workflow

Timesheet Submission & Approval

Monthly Team Check-in & Feedback Loop

5

Operations

Site Induction Process

Construction Project Scheduling (Gantt Chart)

Materials Ordering & Delivery Coordination

Subcontractor Communication & Scheduling

Stage-Based Inspections

Variation Management Process

Practical Completion & Handover Workflow

6

Management

Weekly Leadership Meeting

Quarterly Strategic Planning & Budgeting

Risk Register Update

Team Performance & Productivity Review

Contractor & Supplier Evaluation

Example: Healthcare Services

Minimum Viable Systems™ (MVS)

1 Marketing	2 Sales	3 Finance	4 Human Resources
Client Referral Intake	New Client Onboarding	Batching & Invoicing	Recruitment & Credential Verification
Website & SEO Updates	Initial Client Assessment Scheduling	Claim Submission Process	Onboarding & Orientation
Community Outreach & Events Participation	Funding Verification & Service Agreements	Payment Reconciliation	Clinical Supervision Scheduling
Social Media Scheduling & Management	Client Consent & Intake Documentation	Payroll Processing	Tracking & Policy Compliance
Google Reviews & Testimonial Requests	Program Quoting	Expense Reimbursement Process	Annual Performance Reviews

Weekly Staff Wellbeing Check-In Process

Incident & Feedback Reporting

6

Management

Weekly Leadership Team Meeting

Strategic Planning

Compliance & Accreditation Audit Prep

Systems Development & Review Cadence

Software Procurement Assessment

Aged Receivables Review & Follow-up

Financial Dashboard Reporting

5

Operations

Client Scheduling & Therapist Allocation

Clinical Note Entry & File Audits

Home Visit Safety Check & Protocol

Group Program Delivery Workflow

Interdisciplinary Team Case Review Meetings

Equipment Ordering & Assistive Tech Workflow

Handover Process for Departing Clinicians

Waitlist Management

Conversion Follow-up

Monthly Marketing Metrics & Review

Minimum Viable Systems™ (MVS)

1
Marketing

2
Sales

3
Finance

4
Human Resources

6 Management

5 Operations

Step 3: Share your completed MVS

Once you've worked your way through all the departments, it's time to share what you're doing with the wider team. Here's something I've learned over years of helping businesses systemise: the most successful Systems Champions don't work in isolation. They understand that getting buy-in from the entire team isn't just helpful but essential.

When you share your MVS, several powerful things happen.

First, you get invaluable feedback. The business's team members are in the trenches every day, working with these systems. They might spot critical processes you've missed or help you prioritise what truly matters.

Second, you create alignment. When everyone can see how their department's systems connect with others, they understand their role in the bigger picture. It's like showing each organ in the body how it helps keep the whole organism healthy. This understanding naturally leads to better cooperation between departments.

Third, you build momentum. When team members see their critical processes being documented and valued, they become more engaged in the systemisation journey.

Step 4: Create your MVS scoreboard

Now that you've identified a maximum of 42 critical systems to be documented, shortly you're going to start the process of documentation and organisation. But first, let's create a dashboard to measure your progress. Think of this as your systemisation scoreboard, which everyone can follow – a simple way to track what's done and what's next.

Let's keep it refreshingly simple with a basic spreadsheet. Your scoreboard will track five elements: the department name, system name, current status, knowledgeable person who knows how the task is completed and the target completion date. I made a template for you here: **www.SystemsChampion.com/resources**

If you want to go one step further, you can colour-code the status of each system. White cells show systems not yet started, orange indicates those in progress and green celebrates the ones fully documented.

This visual approach gives everyone on your team an instant snapshot of your progress. There's something deeply satisfying about watching those white cells turn orange, and even more satisfying when they finally turn green. I've seen entire teams get excited about their "green count" growing week by week.

> **PRO TIP:** By treating your MVS tracking like a game everyone can see and join in on, you'll be amazed at how quickly your systems documentation progresses. And there's nothing wrong with a little friendly competition between departments to get things moving.

MVS and CCF: Building on Your Systems Foundation

For the astute reader who has already read my first book *SYSTEMology*, you may recognise the MVS goes hand-in-glove with the concept of the Critical Client Flow (CCF).

The CCF is another tool we use that applies the 80/20 rule. It zeroes in on the core 10–15 systems required to attract, convert and deliver your primary product or service. In effect, it's the most critical 20 percent of your MVS.

The CCF resonated so strongly with readers of SYSTEMology because it offered a clear starting point – a way to break through that initial paralysis of "where do I begin?"

But here's what many missed. The CCF was, and still is, meant to be just that: a starting point. It's a stepping stone on the way to MVS. Your goal is to reach the point at which systemisation touches all departments and all roles within your organisation. If you find it helpful, you can still use the CCF to help you focus on where to start. But don't stop till you reach MVS. Keep your eye on the prize!

11

System for Creating Systems 2.0

A T THE RISK OF AGING myself here ... Did you ever see the TV show *Heroes*? It was about a group of ordinary people who suddenly discovered they had superhuman abilities like time travel, regeneration and telekinesis. As they tried to understand what was happening to them, they found themselves caught in a web of conspiracies, secret organisations and a supervillain named Sylar.

For some reason (and I wasn't sure why until now), the bad guy's method of stealing other characters' abilities always stuck with me. To absorb their skills, he would cut the tops of his victims' heads off, examine their brains and somehow "learn" how their powers worked, effectively absorbing their abilities while killing them in the process.

Sounds pretty gruesome, right? But you can understand why he quickly became so powerful. His real skill wasn't the powers he stole. It was his ability to understand and replicate what made others special.

Now here's the kicker. I'm going to teach you how to do the same thing, but without the killing. Following the last chapter and using the MVS dashboard, you now have a list of the skills you're looking to understand, and you know who you're going to have to ~~kill~~ meet to learn how they do what they do.

I know I'm having a little bit of fun here, but in all seriousness, you don't realise how lucky you are to be doing this role. You're going to be meeting with the best of the best in the organisation and you're going to be learning from them firsthand how they do their role. In time, you will master the skill of skills extraction and repeatability. This ultimate superpower gives you unmatched visibility and understanding of how the business works. This is an extraordinary opportunity, and I want to ensure you realise that.

In my original *SYSTEMology* book, I introduced a system called the System for Creating Systems. It's a step-by-step approach to capturing how people do what they do. It covers everything from identifying the results you're looking to replicate to how processes are recorded, documented and stored. The method was revolutionary in its approach and has helped thousands of businesses build their operations manuals. That said, even till very recently, this has always been the biggest challenge in the whole SYSTEMology process.

I've seen it many times before: business owners get excited about systemising their operations, they identify their critical processes, and then ... everything grinds to a halt at the documentation stage. The traditional approach was time-consuming and costly, and without a dedicated Systems Champion, it was nearly impossible for busy business owners to maintain momentum.

But here's the good news: the emergence of AI has completely transformed how we approach system documentation. What once took hours can now be done in less than a minute. It's almost like magic! A process that previously required multiple drafts and endless revisions can now be completed in a single, efficient workflow. You can now harness AI tools to transcribe recordings, generate initial drafts and even identify potential improvements. All while maintaining the human touch that makes systems truly effective.

What hasn't changed is the fundamental principle. If you're going to create 42 critical systems for your MVS, you need a clear, repeatable process. You need a system for creating your systems. I thought it might just be quicker to give you mine and you can tweak it to make it your own. Introducing the new and improved …

System for Creating Systems 2.0 (Now with Added AI)

Step 1: Identify the result

Start by picking one of the identified systems from your MVS to be your guinea pig. I suggest starting with something simple so you can become familiar with the process. Name it something clear but descriptive, for example, "Posting on Facebook" or "Processing Customer Refunds".

Step 2: Identify who produces the result

Identify the knowledgeable worker who knows how to do the job for which you're creating a system. We'll extract the system from them and then aim to bring everyone else up to that standard.

Planning and communication is key here. I recommend recording a short video (using your preferred screen recording tool) explaining what you're looking to do and send that to the knowledgeable worker, along with some notes on how they can best prepare.

Step 3: Choose your capture method

What's the most effective way to capture your knowledgeable worker in action? Unlike Sylar, we don't need to literally get inside someone's head. A simple video recording will do just fine.

For tasks on the computer, use screen-recording software. For out-in-the-field tasks, consider video recording on your phone. For some processes, an audio interview might work best. The key is choosing the method that creates the least friction for your knowledgeable worker.

Step 4: Record the task

Engineer a comfortable and relaxed environment for the extraction. Remind the knowledgeable worker of the purpose of the exercise and put them at ease by emphasising that this is not about evaluating their process.

Have them walk you through the process, step by step, in the manner in which they would normally perform it, in as much detail as possible. Ask clarifying questions along the way to ensure you understand each step thoroughly.

The word "why" should be foremost in your thoughts throughout. It's not just about the action but also about what the action is intended to achieve. Make a special note of key decision points and potential problems that can arise.

Just remember you're capturing the first version of your system here. Don't overthink this – keep it simple and capture the task as it's being done today. You don't need to cover every variation or every possible combination. Just capture what's most probable.

Step 5: Generate the initial documentation

Here's where our process has evolved significantly in recent times. Instead of spending hours manually transcribing and formatting, we now harness the power of AI to do the heavy lifting. Most modern recording tools include transcription features, so start by getting your recording transcribed.

Then, using AI tools like ChatGPT or systemHUB (more in the next chapter on this), convert that transcription into a structured system document in minutes rather than hours. You'll then review and refine the output to ensure the AI didn't make any obvious mistakes. Look for missing steps and details. Consider organising information for better clarity and logical flow. Consider adding in screenshots and video stills to illustrate some of the steps.

Step 6: Store it in systems management software

Now it's time to store your documented system where the team can easily access it. Whether you're using systemHUB or another platform, you'll need to create the appropriate department folder and save the system with a clear, searchable title.

The goal here is to have a single central location where everything needed to complete this task lives together. And if you haven't already selected your platform, I'll explore this further and provide you with a software checklist in Chapter 13: "Accountable & Transparent".

Step 7: Review with the knowledgeable worker

Once you have your first draft in place, it's time to go back to your knowledgeable worker. Rather than having them simply read through the documentation, ask them to follow the steps the next time they complete the task. This real-world test drive will quickly reveal any gaps, errors or unclear instructions. Remember, they're not just checking for accuracy. They're ensuring the system is useful.

Step 8: Integrate and deploy

With your system documented and reviewed, it's time to hand it over to the team. Work with the department head who'll be overseeing this process. Share the system with them and ensure they understand both the process and the expected outcomes. They're best positioned to train their team and monitor the system's effectiveness day to day.

I suggest they follow this three-step training approach. First demonstrate the task following the system, then complete it together and finally observe them doing it independently. This continues until the team member can successfully complete the task without intervention.

As the system gets used regularly, encourage feedback. The goal isn't perfection from day one: it's continuous improvement over time. Every time someone uses the system is an opportunity to make it a little better.

Work with supervisors to gather insights from their teams and implement valuable improvements.

Remember this: a system is never truly finished. Think of it as a living document that grows and evolves with your business. Your role is to ensure it stays current and effective, while the supervisors handle the day-to-day implementation and training.

Make it your own

I've just walked you through my updated System for Creating Systems 2.0, but here's something important to understand. I've generalised, simplified and added commentary to help with your understanding. I wanted to give you a solid starting point. But remember: every business has its unique ways of working. Your team might prefer different tools, or your industry might require additional steps for compliance or quality control.

So think of my outline like a starting point that you'll build upon and tailor. It's your job to take what I have provided here and turn it into an actual system you and your team can follow.

Would it be helpful to see some more examples of what the final output might look like? Visit: **www.SystemsChampion.com/resources**

12

Extraction Playbook

OKAY, I KNOW THE *HEROES* reference in the previous chapter was a little obscure so I have one more for you. And I promise this will be my last pop culture reference!

Have you read any of the *Harry Potter* books? I've just finished *The Half-Blood Prince*. Please don't judge me. I started reading the series to my son ... Okay, okay, it's true, I'm kind of enjoying it too now.

Anyway, there's this moment in the book where Harry stumbles across an old, battered potions textbook. It's filled with scribbled notes, crossed-out instructions and tiny tweaks to the recipes. Shortcuts that turn him into a potions master.

Instead of following the textbook's steps to "Cut the sopophorous bean and add it to the cauldron," Harry's version says to "Crush the sopophorous bean with the flat side of a silver dagger to release the juice more effectively." These simple tweaks make his potion far more effective than everyone else's, including Hermione's (which, if you've read *Harry Potter*, you know is a big deal). Harry shoots to the top of the class, catching the attention of Professor Slughorn.

To cut a long story short, it turns out the book was once owned by the mysterious Half-Blood Prince, who spent years perfecting those recipes. Now, I don't want to give any spoilers, but I *would* like to offer you the same advantage given to Harry.

Let me be your Half-Blood Prince of systems documentation.

As you can imagine, over the years, I've collected plenty of helpful tips and shortcuts that can dramatically improve how you document systems. Surprisingly, many common assumptions about creating systems turn out to be wrong.

For instance, many believe systems should be comprehensive, documenting every possible scenario – yet the best systems are often the simplest. Others think documentation should be formal and technical – but systems written in a more conversational language get used more often. And contrary to popular belief, writing systems for the lowest skill level can actually hurt your business by driving away your most talented team members.

What follows is a collection of battle-tested lessons that might challenge what you think you know about systemisation. Each one addresses a specific challenge you're likely to face. Whether you're dealing with resistant team members, struggling with the level of detail to include or trying to decide what's worth documenting, you'll find guidance here.

So, just like Harry with his potions book, you now have access to my notes that will make extractions, documentation and systemisation faster, easier and way more effective. Let's get started.

Lesson #1: Create only useful systems

Arguably one of my most important rules of extraction is to ensure you're creating systems that add value. Avoid the trap of creating systems for systems' sake. You already reduce that chance of misstepping here by sticking to the MVS, but beyond that remember this: not everything needs documenting.

Before you invest time, ask yourself one crucial question: "Will this documentation genuinely help the business?" If you can't clearly see how this system will enhance customer experience, free up more skilled team members, improve efficiency, reduce errors or deliver some other tangible benefit, pause and reconsider.

Every system you create requires time to extract, document and maintain. Time is money, and that's time you could spend on other, more valuable systems. Remember that the documentation is not the goal; the goal is to improve the business!

Lesson #2: Master the art of extraction

There is a direct correlation between how well you handle the extraction process and the quality of the documented system you produce.

Ever heard the computing expression "garbage in, garbage out"? It perfectly describes what happens when you don't get this part right. If your recording is poor quality, if the knowledgeable worker rushes through the steps or if you miss crucial details during extraction, your final system will be limited in its usefulness. Or worse, it might introduce errors into the process.

This is especially true now that you're using AI to help create your documentation. AI works best when you provide clear, structured input. Speaking clearly, marking transitions between steps ("That completes step three, now let's move on to step four") and maintaining a logical flow makes a massive difference. The better organised your extraction is, the better job AI will do at creating your initial draft. Think of AI as your documentation assistant – the clearer your instructions, the better its output.

Yes, the knowledgeable worker is the source of the process, but you are the driving force behind its successful extraction. You have the power to ask the right questions, uncover hidden gems of knowledge and ensure nothing is lost in translation.

PRO TIP: Here are two simple tools that can be used before any extraction to dramatically improve your results.

#1: Team introduction video: Create a short video (under 10 minutes) explaining why the business is investing in systems and how it benefits everyone. Introduce yourself as the Systems Champion and explain how you'll support the team through the process. Think of this video like a movie trailer – it sets expectations and gets people excited about what's coming. Keep it simple and authentic.

#2: Pre-extraction form: Send your knowledgeable worker a brief questionnaire to help them organise their thoughts. Ask them to note down what tools or logins they'll need, the main steps at a high level and any common pitfalls to watch out for. This brief preparation helps them organise their thoughts and often reminds them of important details they might otherwise forget in the moment.

Remember, every minute you invest in getting a quality extraction will save you multiple minutes in documentation and revisions later. Taking time to prepare, both yourself and your knowledgeable worker, is always worth the effort.

Extraction Checklist

Before recording:

- ❏ Create introduction video
- ❏ Send pre-extraction form
- ❏ Test recording equipment
- ❏ Book quiet space
- ❏ Prepare key questions
- ❏ Arrange backup method

During recording:

- ❏ Check audio quality
- ❏ Note decision points
- ❏ Mark critical steps
- ❏ Ask "why" questions
- ❏ Document potential issues

After recording:

- ❏ Transcribe audio
- ❏ Have AI draft process
- ❏ Add screenshots
- ❏ Get expert review
- ❏ Identify gaps
- ❏ Note improvements

Lesson #3: Capture what's working

When you start documenting a process, it's tempting to immediately look for ways to improve it. Your analytical brain kicks in, and suddenly you're seeing dozens of potential optimisations. While that innovative thinking is valuable, it's not what you need right now.

Your goal is to capture what's already working in your business today, focusing on the most probable path. That is, what's most likely to happen. We're looking for the path that represents the "blue sky scenario" where everything goes according to plan.

You don't have to document every possibility. When exceptions occur (and they're not accounted for in the system), team members will escalate things to more experienced members of the team. Your documented systems allow newer team members to master the core process without getting overwhelmed by every possible variation.

By focusing on what's currently working and the most probable path, you create several crucial benefits. Firstly, it creates a baseline: after all, you can't measure improvement without knowing where you started. Secondly, it keeps things moving forward. The moment you start trying to optimise processes or account for every exception, everything tends to slow down. Thirdly, and perhaps most importantly, it makes training easier because new team members can learn the core process first, then gradually understand how to handle exceptions as they gain experience.

Remember: right now, your job is to create consistency. Get everyone following the same main path, even if it's not perfect. Once you have that consistency, you'll have the foundation needed for meaningful improvement.

Lesson #4: Choose the right system type

As you begin extracting systems, you'll notice they generally fall into two categories: higher level overview systems and more detailed "how-to" systems. Understanding the difference helps you choose the right documentation approach from the start.

Overview systems are exactly what they sound like: high-level processes that give you the big picture. Think of creating an annual marketing plan or managing a large project. These systems often span longer timeframes and involve multiple steps that can't be completed in a single sitting. When documenting these, you're usually having a discussion rather than recording someone performing a task in the moment.

For example, if you need to document the way your primary product or service is delivered. There's a chance it could be quite complex with many moving parts. That might be an overwhelming place to start. Instead of jumping right into the detail, you might start with an overview. You'd sit down with your knowledgeable worker and have them walk you through just the key milestones at a high level.

How-to systems, on the other hand, are your detailed, task-specific processes. These are the day-to-day operations like issuing invoices in QuickBooks or updating your CRM. For these systems, it's best to record the actual process being performed, capturing every click and keystroke. This is where you need that step-by-step precision because these systems are often used to train new team members or serve as reference guides for tasks that might only be done occasionally.

Here's my hot tip when recording detailed how-to systems: slow everything down. Your knowledgeable worker probably performs these tasks on autopilot, which means they might zip through important steps without explanation. Encourage them to move deliberately and narrate their actions, and be sure to ask lots of questions if they're not clear. These seemingly obvious details often make the difference between a useful system and one that leaves new team members confused.

The key is being flexible in your approach. Sometimes you'll need to record a process in action, and other times you'll need to extract it through discussion. Let the nature of the system (and who will be using it) guide your documentation method. And when in doubt, start with a high-level overview and look to add more detailed "how-to" systems over time.

Lesson #5: Balance detail with usability

"How detailed should my system be?" There's no perfect answer to this question. It depends. What works beautifully for one process might be completely wrong for another. But there are three key factors that will guide your decision:

1. Who's using it? A highly skilled team member with years of experience might only need a high-level checklist, while a new hire will need more step-by-step guidance. For example, your experienced bookkeeper might just need bullet points for month-end reporting, but your new admin assistant will need detailed screenshots for processing invoices.

2. What's the complexity of the task? Simple, straightforward tasks might only need a few bullet points and a video. Complex processes often benefit from being broken into smaller chunks.

3. How often is it used? Tasks performed daily might need less detail as people quickly become familiar with them. But processes done quarterly or annually? Those need more detail because people don't have the benefit of regular practice to reinforce their learning.

The best systems are like a good pair of jeans. They fit well in most situations and get better with use. If you find yourself writing a system that only works under perfect conditions with a specific person at a specific time, you've probably gone too far into the details. Think about it. The more specific and rigid your system becomes, the more likely it is to break when circumstances change even slightly.

Keep your systems flexible enough to handle most situations yet specific enough to still deliver consistent results. You want to create a framework that guides people to success, not a straitjacket that restricts their ability to think and adapt.

Remember, you can always add more detail later. Start with what's necessary to get the job done correctly, then refine based on feedback and actual use.

Lesson #6: Make systems human-friendly

You're writing for humans, not robots (at least, for now). Your systems need to feel like a helpful colleague guiding someone through a task, not a cold technical manual dictating commands.

This starts with language. Write conversationally, as if you're explaining the process to a colleague over coffee. Instead of "Initiate the customer complaint resolution protocol," write "Here's how to handle a customer complaint." Skip the corporate jargon and speak in terms your team actually uses.

You might not realise it but while you're writing for humans, this approach works better for AI too. Clear, conversational language that's well-structured and logical helps both people and machines understand your processes. It's a win-win. Make it clear enough for a person to follow, and you've probably made it clearer for AI to process too.

Put yourself in the shoes of a new hire using your system for the first time. They probably don't know your internal acronyms or have years of experience doing this task. What seems obvious to you might be completely foreign to them. This perspective helps you identify what really needs to be explained and what can be assumed.

Just be mindful to keep it engaging. Long, dull systems are like long, dull meetings – people find ways to avoid them. If a system feels too lengthy, consider having an overview system and then breaking up the detail into smaller detailed how-to systems.

Lesson #7: Simplify everything

Einstein said we should make things "as simple as possible, but not simpler". He could have been talking about business systems.

Most people overcomplicate their systems, thinking they need elaborate templates and rigid frameworks. The reality is much simpler. As long as your approach is consistent and helps people get the job done, you're on the right track.

Most effective systems include these key elements:

- **Title:** A clear heading that tells people exactly what this system achieves. Make it descriptive and searchable. I'll often start with a verb, like "Processing Customer Refunds" or "Creating a YouTube Thumbnail".

- **Purpose:** A statement that explains the "why" behind the process. This helps people understand the value of what they're doing and make better decisions along the way. For example, "This system ensures customers receive prompt refunds while maintaining accurate financial records."

- **Trigger:** An explanation of what kicks this system into action. Is it a customer complaint? A monthly deadline? A specific request? Make it clear when this system should be used.

- **Inputs:** What you need to have ready before you start. This might include information, documents, tools or access to specific software. Nothing is more frustrating than getting halfway through a process and discovering you're missing something essential.

- **Steps:** The actual process, laid out in a logical sequence. This is your meat and potatoes of the system. What needs to happen from start to finish? Break it down into manageable chunks.

- **Endpoint:** How people know when they're done. What does success look like? How do they know they've completed the process correctly? Define your outputs and what happens next. Does another system kick in? Does someone need to be notified? Make the hand-off points clear.

▪ **Resources and examples:** Any templates, tools or links needed along the way, plus real-world examples of the system in action. Nothing beats seeing how something works in a real situation.

That's it. No need to overcomplicate things. Find a format that works for your team and stick with it. The goal is to make your systems useful, not win any documentation awards.

Lesson #8: Update when needed, not scheduled

I remember working with a business owner who proudly told me they review all their systems every December. When I asked why December, they paused and realised it was simply because that's what they had always done. Like many businesses, they had fallen into treating system updates like an annual tax return – something to check off once a year and forget about.

This "review everything annually" mindset often comes from ISO certification requirements or corporate policies. But forcing updates on a rigid schedule usually leads to one of two problems: either it becomes a meaningless checkbox exercise, or worse, people wait to fix broken systems because "the annual review is coming up anyway".

Think of it like maintaining your car. You don't wait for an annual service to fix a flat tyre, right? You fix it when it's broken. The same goes for your systems. Update them when they need updating.

I want to make sure an important distinction is clear. Updating a system is not the same as re-engineering it. Updating means tweaking and improving what is there, clarifying steps, adding missing information or reflecting minor changes in how things are done. Re-engineering is about fundamentally rebuilding the process from the ground up, which is a different exercise entirely.

I will talk a little more about re-engineering and optimising systems later, but for now, the key is creating a culture where everyone feels empowered to flag when a system needs updating. Your team members are your best early warning system. They know when something is not working well.

Putting it all together

There you have it: some practical tips drawn from years of helping businesses just like yours capture their way of doing things.

Remember, your goal isn't to create perfect systems. It's to capture what's working in your business today, make it repeatable and build a foundation for continuous improvement. Start with clear recordings, focus on what matters most and keep your systems human.

You might be surprised how quickly things start falling into place once you begin. That first system might take a bit longer as you find your rhythm, but by system three or four, you'll have developed your own style and the process will feel natural.

I've seen this transformation happen countless times. Business owners and their Systems Champions who once struggled with systemisation are leveraging these ideas and AI and suddenly find themselves with a growing library of practical, usable systems. Their teams become more confident, their operations more consistent and their businesses more scalable. Most importantly, they've created a culture where "I didn't know how" becomes "Let me check the system."

Now it's your turn. Pick your first system, schedule that extraction and start building.

Your First System Worksheet

Selected process: ....

Process frequency:

❏ Daily ❏ Weekly ❏ Monthly ❏ As needed

Why did you choose this process?

❏ It's simple and straightforward

❏ It's frequently performed

❏ It has a clear outcome

❏ It's one the team is comfortable sharing

❏ Other: ....

Expected outcome: What does success look like for this process?

....

....

Knowledgeable team member: ....

Best way to approach them:

❏ Direct conversation

❏ Email introduction

❏ Through their supervisor

❏ Informal chat

I commit to having documented my first system by (date):

....

Case Study   From Interior Designer to Systems Champion

When Ryan Stannard grew his carpentry business into a $15 million custom home-building operation in Adelaide, he faced a common entrepreneur's dilemma: he was trapped by his own success. He found himself answering endless questions and unable to step away from the business. He needed someone to help systemise operations, but the solution came from an unexpected source – his daughter Eryn.

Like many business owners, Ryan realised that to scale, he needed to implement robust systems and processes. "I needed to clone myself," he recalls, recognising that without proper documentation, the business would always depend on him.

Enter the Systems Champion

When Eryn joined Stannard Homes, she brought a passion for interior design, not systems. In fact, her tendency to question everything led to the existing interior designer quitting within her first four weeks. But this questioning nature would prove invaluable for the business's transformation.

Ryan shared the company's systemHUB account with Eryn, which contained their documented processes. But instead of simply following these systems, she began questioning, improving and rebuilding them. Within six months, she had completely rewritten the interior design systems and was handling 12 client selections simultaneously.

What made Eryn particularly effective was not just her ability to document processes. It was her curiosity about how everything connected. Every system she wrote or updated deepened her understanding of the business. She moved from interior design to assistant manager, gradually becoming one of the most knowledgeable people in the company.

"She now knows every intricate bit of the business because she's been rewriting the systems manual," Ryan explains.

"She's not just documenting processes. She's understanding how everything fits together."

The results of systematic change

The impact has been transformative. Stannard Homes has doubled in size, growing from seven to 15 staff members. Ryan can now take extended holidays knowing the business will run smoothly. Most importantly, as Ryan prepares to launch a new venture in a rural area, Eryn is ready to manage the existing $15–20 million operation.

Ryan believes having a Systems Champion is a "gift" to the business. "You don't need someone with documentation experience," he advises. "You need someone who's open-minded, a good listener and a 'go-getter'. The Systems Champion role can be a viable career path for ambitious employees."

Through the combination of a business owner willing to let go and a Systems Champion eager to learn and improve, Stannard Homes transformed into a scalable, systematic operation ready for its next phase of growth.

Watch the full interview here:

www.SystemsChampion.com/resources

Pillar 1 Action Items

☐ Document your Minimum Viable Systems (MVS). Work through each department to identify and list your critical seven systems per department, creating your MVS dashboard to track progress.

☐ Customise and document your own version of the System for Creating Systems 2.0. Adapt the framework to fit your business's unique needs while maintaining its core principles.

☐ Create your pre-extraction toolkit: prepare your introduction video, design your pre-extraction questionnaire and select your recording software.

☐ Choose and document your first test system. Pick something simple but important to help you master the documentation process while building momentum.

☐ Schedule regular check-ins to review your progress. Block out time each week to ensure your documentation efforts stay on track and maintain momentum.

PILLAR 2: TOOLS

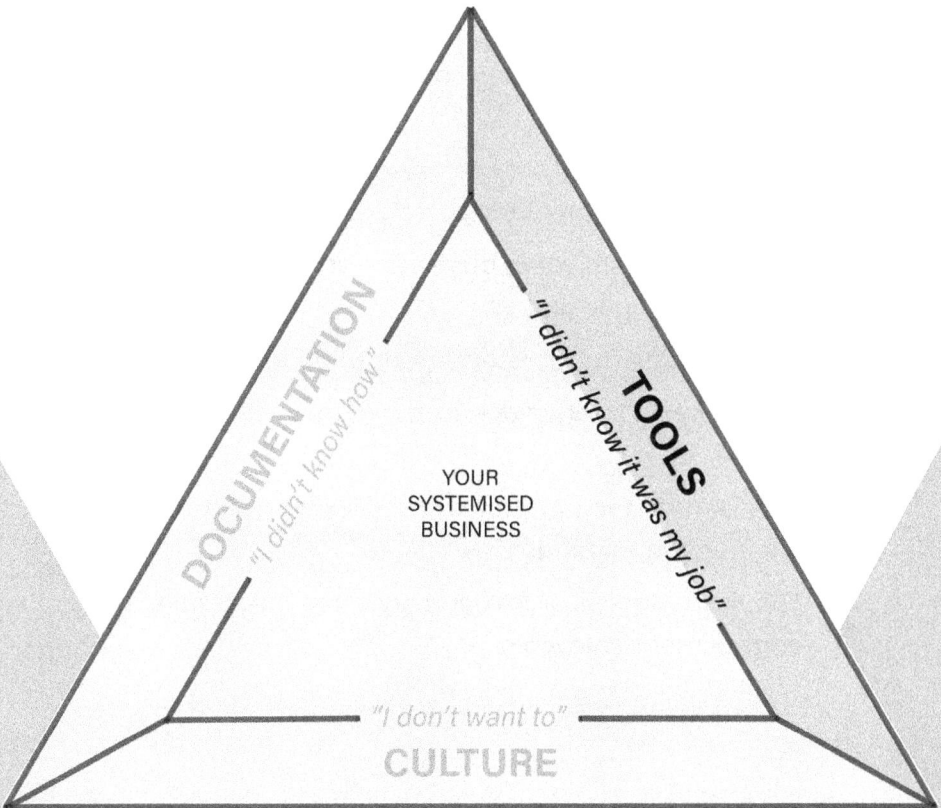

Inside the triangle diagram:

- DOCUMENTATION — "I didn't know how"
- TOOLS — "I didn't know it was my job"
- CULTURE — "I don't want to"
- YOUR SYSTEMISED BUSINESS (center)

"Tell me how you measure me, and I will tell you how I will behave."

Eliyahu M. Goldratt, creator of the Theory of Constraints and author of The Goal

Summary

Building a systems-driven culture requires more than just documentation. It demands the right tools, training, transparency and accountability. Success comes from creating an environment where systems are easily accessible, consistently used and supported by modern technology including AI.

Highlights covered in these chapters include:

- Why transparency and accountability are fundamental to systems success.

- The critical balance between systems management and project management tools.

- How AI is transforming business systemisation and creating new opportunities.

- The evolution from Systems Champion to AI Champion.

- The power of workshops in building system adoption and habits.

- Why regular, focused training sessions outperform one-off intensive training.

- The importance of making systems accessible and engagement enjoyable.

13

Accountable & Transparent

IN 1989, MICHAEL JORDAN WAS already being called the greatest basketball player of all time. His aerial acrobatics and scoring ability were unprecedented. He was putting up incredible numbers. In one season he was averaging 37 points per game. But despite Jordan's individual brilliance, the Chicago Bulls weren't winning championships.

Everything changed when Phil Jackson became head coach. He introduced a new offensive system called the "triangle offense" that forced Jordan to trust his teammates more. At first, Jordan resisted. He'd grown used to taking on entire teams by himself. But Jackson convinced him that to reach the next level, he needed to empower his team.

What happened next was extraordinary to watch. Jordan started making different plays. He'd drive toward the basket, drawing defenders to him, then suddenly whip the ball to Scottie Pippen in the corner or fire a pass to John Paxson at the three-point line. These weren't gentle passes; they were bullets, thrown with absolute conviction that his teammates would be exactly where they needed to be, ready to catch and shoot. And they were.

Because the triangle offense wasn't just a collection of plays. It was a tool that created absolute clarity about where everyone should be and what they should be doing. It gave Jordan new options he'd never had before. He could now make no-look passes to Horace Grant under the basket because he knew, without even looking, exactly where Grant would be.

The results? Six NBA championships in eight years. Jordan didn't just maintain his greatness: he elevated it. His scoring might have dropped slightly, but his assists went up, his teammates' production soared and, most importantly, his team started winning at an unprecedented level.

Just like basketball, business is the ultimate team sport. You might have incredible talent in your organisation, but true success comes from getting everyone working together seamlessly. I see it all the time – businesses struggling not because they lack talent, but because they're playing like a pre-1989 Bulls team. They have their Michael Jordans, but they're trying to win championships through individual brilliance alone.

What makes a poorly performing team in the game of business? People working in silos, keeping their processes hidden in their own little black boxes. Sometimes it's deliberate – they think keeping knowledge to themselves provides job security. Other times, it's simply because it's "how we've always done it". Either way, these hidden pockets of information hurt the entire team's performance.

By the nature of the task, documenting your systems will shine a light on this, making the invisible visible. But it is the tools that will truly create transparency and accountability, ensuring everyone knows exactly who's doing what and by when. This clarity is what turns a collection of individual stars into champion-winning teams.

But before you think I'm going to give you specific tool or software recommendations, you can relax. In truth, those are just the details. And in reality, which tool you select is not what's important here. Different industries use different tools, and different businesses and their teams will choose the software that is right for them.

What I am going to share goes deeper. We are going to look at what really needs to be in place to have teams operating at their highest level.

Because when you get this right, when everyone has clarity about who is doing what, that is when business becomes a beautiful team sport.

The two key areas to focus on when you're selecting your tools are transparency and accountability.

Think of transparency like putting your whole team in a glass-walled office. Everyone can see what everyone else is working on. There is nothing to hide, no secret projects, no mysterious priorities – just clarity about who is doing what and when it needs to be done. This visibility is not about micromanagement. It is about empowerment. When team members can see how their work connects to others, they naturally start finding ways to improve the whole system.

This transparency leads naturally to accountability. When work is visible, it automatically drives people to take ownership. Every team member becomes responsible for completing tasks, and it quickly becomes obvious if someone is letting the team down.

The magic happens at the intersection of documented processes and the right tools. Together, they eliminate those all-too-familiar excuses: "I didn't know it was my job," "I didn't realise it had to be done by then," or the classic "Oh, I forgot." When expectations are visible to everyone, these excuses simply vanish. Clarity leaves no room for ambiguity.

So, what tools do you need to create transparency and accountability?

There are countless tools out there promising to transform your business. But rather than getting lost in a sea of software options, let's focus on the two types that matter most for systems-driven businesses: systems management and project management.

Systems management

You may have already given systems management software some thought when you worked through the previous section on documentation. This is simply the space where your documented systems are stored. It's the central repository for the "how-to" knowledge within the business. It should be simple, centralised and easily accessible.

The secret to solving the "I didn't know how" excuse is to ensure team members are never more than one click away from the process they need. When someone's in the middle of a task and needs guidance, they shouldn't have to hunt through folders or remember complex navigation paths. Every system should be instantly accessible right when and where it's needed.

This might mean embedding links in task descriptions, sticking QR codes near equipment that links directly to the associated process or including direct links in calendar invites. The goal is to remove any friction between your team members and the information they need. When someone says "I couldn't find the system," that's not their failure – it's a sign your set-up needs improvement.

Most people get this wrong. You don't need the fanciest software with the most features. In fact, chasing features often leads to complexity, and complexity is the enemy of adoption. What you need is simplicity and accessibility. Remember, the best system in the world is useless if your team can't access it exactly when they need it.

One of the topics I have received some mixed feedback on was my recommendation in *SYSTEMology* to keep your systems management software separate from your project management software (which we'll talk about shortly). And while there have been advancements with different tools, I still believe there are several compelling reasons to keep your systems documentation separate.

First, your documented systems are arguably your most valuable business asset. They capture years of experience, refinement and proven success. Housing them in dedicated software sends a clear message about their importance and ensures they receive the attention they deserve. It's like keeping your family's treasured recipes in a proper cookbook rather than scattered among grocery lists and Post-it notes.

Second, it establishes a single source of truth. When team members need to know how to do something, they know exactly where to look. No more hunting through project boards or wondering if they're looking at the latest version. Everything is in one dedicated place, always current, always accessible.

Third, it protects and preserves your intellectual property. While project management tools are designed for temporary, task-focused information that comes and goes, your systems are permanent, valuable assets that need proper protection and management. When processes evolve (and they will), you can update them in one place, knowing that every linked reference points to the current version.

Think of it like this: your project management tool is your daily to-do list, while your systems management software is your business's encyclopedia. They serve different purposes, and keeping them separate makes both more effective.

When selecting your systems management software, there are a few features I recommend looking for:

Systems Management Software Criteria

- Dedicated systems management platform: Look for software specifically designed for creating and organising business systems, not generic document storage platforms or wikis. Your most valuable business asset deserves a proper home.

- Rich media integration: Make sure the platform allows you to embed videos, images, attachments and other resources directly within your systems. Everything related to a process should be accessible in one place.

- Permission controls: You'll need varying levels of access for different team members. The ability to assign systems to specific roles then assign those roles to individuals is crucial for security and organisation.

- Accountability features: Look for features that allow team members to sign off on systems they've read and understood. This eliminates the "I didn't know" excuse and creates clear accountability.

- Intuitive user experience: The best system in the world is useless if it's too complicated to use. Your platform should require minimal training and feel familiar to team members with varying technical abilities. What you want is a great pocketknife, not a Swiss Army knife with 50 tools you'll never use.

Remember, complexity is the enemy of adoption. There's no need to overthink this. See our current recommendations at

www.SystemsChampion.com/resources

Project management

Once you have your systems management software in place, it's time to tackle the second piece of the puzzle: project management. That is, how you will organise who is doing what by when.

If you're already using project management software or some other tool to manage your tasks that you're happy with, great. You can skip ahead to fine-tune your approach. But if not, I want to prepare you. This is going to be a significant change in how your business operates. It's not just about adopting new software; it's about transforming how your team communicates and coordinates work.

Here's the thing, though. If you want to eliminate the "I didn't know it was my job" excuse, project management software isn't optional. It's essential. You need a central place where everyone can see who's doing what and when it needs to be done. Email chains, verbal instructions and sticky notes just won't cut it anymore.

But timing is everything. If you've just started your systemisation journey, focus on your systems management first. Get your "how-to" knowledge documented and organised. Once that foundation is in place, you'll be ready to tackle project management software, and your team will be more receptive to the change because they've already experienced the benefits of good systems.

When you're ready to make the switch, good project management software should do several essential things. First and foremost, it needs to make task ownership crystal clear. There should never be any doubt about who's responsible for what. It must allow you to set explicit deadlines and priorities so everyone knows what needs to be done first and when it's due. All task-related communication should stay in one place, eliminating those endless email chains and scattered conversations.

It should seamlessly link to your documented systems, so team members are never more than a click away from the "how-to" guides they need. You don't need anything fancy – just to be able to place a link in the task description.

And finally, it should make it easy to track progress, so everyone can see how projects are moving forward and where things might be getting stuck. This is what creates transparency and accountability.

Your documented systems are the "how" of your business, while project management is the "who", "what" and "when". Together, they go a long way to eliminating the two biggest excuses that hold businesses back: "I didn't know how" and "I didn't know it was my job."

Setting up your software

Let's briefly talk about how to organise your documented systems. You want to organise everything so there's a logical structure, making it easy to find things when they're needed. The simplest approach is to create folders in your systems management software that match your core departments, e.g. Marketing, Sales, Operations, Finance, HR and Management. You can then use subfolders underneath those, if needed, for further sorting.

It may look a little bit like this:

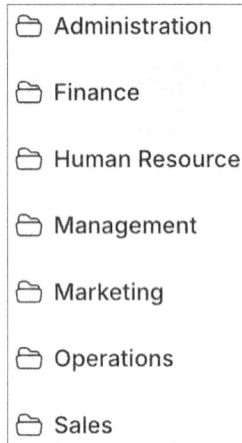

🗁 Administration

🗁 Finance

🗁 Human Resource

🗁 Management

🗁 Marketing

🗁 Operations

🗁 Sales

That looks easy enough! Now, let's combine it with your project management software. Building on this same foundation, we can add one more dimension: time.

Obviously every project management tool is a little different, but try to replicate the same departmental structure with separate areas for each department. Next, within those areas, look to organise tasks by their cadence (i.e. triggered, daily, weekly, monthly, quarterly, annually).

This is the same thinking we used earlier when we completed your MVS. Think of it like creating a rhythm for your business. All you're looking to do is transfer that into your project management software. Here's an example of what it might look like in your marketing department:

Marketing Department (Project Management Software Set-up)

Triggered tasks:

- New lead follow-up

- Social media comment responses

- Website enquiry handling

Daily tasks:

- Social media engagement
- Email inbox management
- Lead qualification

Weekly tasks:

- Performance metrics review
- Content calendar updates
- Team huddles

Monthly tasks:

- Campaign reporting
- Budget review
- Strategy adjustments

Quarterly tasks:

- Major campaign planning

Annual tasks:

- Strategic planning
- Year in review

Clearly I've gone a step beyond the prescribed seven systems per department as part of the MVS, but I wanted the example to be crystal clear. By organising tasks this way, you create natural cycles of work. Be sure to assign the team member who is responsible for each task so everyone knows what's coming up. No more tasks falling through the cracks or last-minute scrambles to meet deadlines.

And here's where everything connects beautifully. Each task in your project management software should link directly to its corresponding system in your systems management software. For instance, when some-one gets assigned "New Lead Follow-Up", they should be able to click straight through to the documented process explaining exactly how to handle new leads.

Remember, you don't need to build this all at once. Pick a department you feel would be easiest to start with and make it your ideal department prototype. Start with the essential tasks you identified in your MVS work. Focus on getting those core processes flowing smoothly before adding more departments, tasks and further complexity.

Here's how you know it's working

Now, I make no apologies for the difficulty in rolling out this set-up in your company – especially the project management software. It's hard work but it's worth it.

At first, you'll start noticing little changes. Team members stop asking the same questions repeatedly because they know where to go to find answers. More tasks get completed without prompting because ownership is clear. Deadlines are met more consistently because everyone can see what's coming up.

But the real magic happens when you notice the business owner feels comfortable stepping back. Not because they have to, but because they can. Through your work in documenting every process and making every task visible, you've created unprecedented clarity in your business operations. The business owner can take comfort knowing they could dive deep into the details if they wanted to, but they don't have to.

Ready to make this a reality? Keeping documenting those processes and making them easily accessible. When that foundation is solid and the time is right, install project management software to create ultimate transparency and accountability. Remember, you're not just organising tasks. You're transforming how your team works together.

Your whole team, including your business owner, will thank you.

Tools and Transparency Assessment

What's holding your team back right now? (Check all that apply)

❑ Unclear task ownership

❑ Siloed work processes

❑ Inaccessible documentation

❑ Missed tasks

❑ Missed deadlines

❑ Tool overload/fragmentation

❑ Other: ....

Systems storage and access

Systems storage method: ....

Is it working well? Yes / No

If no, what's the biggest issue?

❑ Poor searchability

❑ Lack of centralisation

❑ Outdated content

❑ Access restrictions

❑ Other: ....

Three ideas to make systems more accessible:

1 ....

....

2 ....

....

3 ....

....

Project management (PM)

Do you have PM software? Yes / No

If yes, what are you using? ....

Is it working well? Yes / No

Who would be best to chat with about this?

....

14.

In the Field

'M GOING TO KEEP THIS section brief for a couple of reasons. Firstly, the tools landscape changes rapidly with new releases launched daily, and secondly, I don't think it's the best use of your time to go chasing shiny objects. In the early days especially, it's easy to get caught testing out the latest and greatest gadgets feeling like you're being productive. But trust me, this isn't the best approach.

You're best off mastering a handful and keeping it simple. In the previous chapter, I talked about the two foundational tools you'll need: your systems management and project management software. Now let's talk about the other core tools you're going to need specifically for your documentation efforts. There's recording, storage, transcription and some general office suite tools.

It might sound like a lot, but the good news is, you're probably already familiar with some and have access to others. Let's break down exactly what you need to get started ...

Video recording

Perhaps your next most essential tool will be your video recording tool. Whether you're documenting a complex software process or capturing a physical procedure, being able to record what your knowledgeable workers do is crucial.

For screen-based processes, you'll need reliable screen-recording software. The good news? Most computers already come with this functionality built in. And if not, there are plenty of screen recording solutions available, from basic to advanced options. Who knows, you might even be using a fancy AI assistant notetaker that will do the job.

For physical processes (those tasks that happen away from the computer) a smartphone is often all you need. Your phone captures high-quality video that's more than adequate for process documentation. Whether you're recording warehouse procedures, customer service interactions or manufacturing processes, just keep it simple.

I remember working with Kane, a Systems Champion at PorterVac, a company that cleans roofing gutters. Kane faced an interesting challenge: how do you document processes that happen two stories up in the air when you don't have any hands free?

His solution? A GoPro camera attached to his head. He'd follow the tradespeople around, recording their every move. From their initial safety checks to their conversations with clients, from how they set up their equipment to how they sent their work back to head office after a job was complete. Nothing fancy, just a raw video recording that got the job done.

Just keep in mind that clear audio is just as, if not more, important than the video image itself. This audio not only ensures the viewer can follow along but will also play a key role in producing transcripts that you'll feed into AI for documentation. Therefore, if your audio quality is poor, it's worth the investment to get a quality microphone.

Video storage

Next up, you'll need somewhere to store your video recordings. While some screen-capture software might handle this for you and may even allow for additional video uploads, you'll likely still need a dedicated video-hosting solution to ensure smooth playback. Nothing kills your flow quite like waiting for a training video to load!

Professional video-hosting platforms like Wistia, Vimeo or even YouTube can be excellent solutions. Now, I know what you might be thinking: *Can't I just store videos on our shared drive, like Google Drive or SharePoint?* While technically possible, dedicated video hosting offers several advantages, from smoother playback across all devices to no massive storage files clogging up your drive. These platforms handle all the technical heavy lifting of video compression and delivery.

If budget is a concern, start with YouTube's private video options. You can always upgrade to a more business-focused platform as your needs grow. Just ensure whatever platform you choose allows for embedding videos directly in your systems management software. Your team shouldn't need to leave your systems platform or download files to watch process videos.

Transcribing

Now that you've captured and stored your videos, let's talk about turning them into usable documentation. Converting video content into written documentation is a crucial part of your systems work. The good news is, there's a good chance one of the recording or storage tools you select will have this capability built right in.

Just learn which tools do and don't, and think about how to seamlessly integrate transcribing your videos into your documentation workflow. Rather than seeing it as a separate step, think of it as part of your natural process. Record your video, generate the transcript, feed the output into AI and shape that content into clear, usable documentation.

Office suite essentials

Whether you're using Microsoft Office, Google Workspace or another platform, these everyday tools play a crucial supporting role in your systems work.

Your calendar becomes your best friend for coordinating extraction sessions and team workshops. Email (if you're not using internal communication tools) helps you keep everyone in the loop about systems progress. And cloud storage gives you a place for those miscellaneous but important files that don't quite fit in your systems or project management platforms (e.g. working documents, spreadsheets, presentations, videos).

Think of your office suite as your catch-all solution: it fills the gaps between your more specialised tools. While it might not be the star of the show, you're using this tool daily. Just make sure you're familiar with the basic features and shared drive locations your team uses.

> **PRO TIP:** If you're in a position to make suggestions about the way things are set up, wherever possible, try to match your department structure across multiple tools. For example, in your cloud storage, create folders for each department (Marketing, Sales, Operations, Finance, etc.) and organise your files in those accordingly. This consistency among your tools will go a long way to helping your team find things when they need them most. Consistency always wins.

But wait, there's more!

So far, I've focused primarily on the tools you'll need for the job of the Systems Champion. However, it's worth noting you're going to become something of a software anthropologist. Through your work, you'll discover tools you never knew existed in your organisation and you'll learn to understand how they all fit together.

You'll find each department has its own set of tools – its own digital territory. Accounting might live in Xero or QuickBooks. Marketing probably has a suite of tools for social media and email campaigns. Sales likely has their CRM. Your job isn't to become an expert in all of these, but you do need to understand their purpose and how information flows between them.

I'll give you some direction on taking a software inventory in the "Implementation" chapter toward the end of the book, but I did want to mention it here. This knowledge becomes invaluable when you're documenting processes that cross department boundaries. It helps you understand the full journey of information through your business. Plus, you'll know exactly who to talk to when you need deeper insights into specific tools.

Your job isn't to become an expert at every tool within the business. In fact, it's best to be tool agnostic and try not to become overly dependent on any single piece of software. Tools change, each department does things differently and new solutions emerge. Focus on the principles and practices that make things work, not the specific tools you use to achieve it.

Start with recording, storage, transcription and your office suite. In the next chapter, we'll explore where AI will fit into your toolkit. Find out what you already have access to, focus on those core tools and if you need more help, check out the recommendations at:

www.SystemsChampion.com/resources

Your "In-the-Field" Toolkit

Let's map out the essential tools you'll be using.

Screen-recording tool: ....

Physical process-recording tool: ....

Video-storage platform: ....

Transcription tool: ....

Office suite platform: ....

15

AI Champion

IN 2000, BLOCKBUSTER WAS THE undisputed king of home entertainment. With over 9000 stores worldwide, they'd made movie rentals a part of everyday life. Their brand was so powerful that when a small start-up called Netflix offered to sell themselves to Blockbuster for $50 million, Blockbuster's CEO practically laughed them out of the room.

We all know how this story ends. Netflix is now worth many hundreds of billions and Blockbuster ... well, there is exactly one Blockbuster store left in the entire world and it's more of a tourist attraction than anything else.

What happened? Blockbuster failed to recognise a fundamental shift in how people wanted to consume entertainment. While they were busy protecting their late-fee revenue and expanding their physical stores, Netflix was betting on a digital future. By the time Blockbuster realised streaming was the future, it was too late. They'd missed their moment.

It's a cautionary tale we've seen play out many times. Kodak invented the digital camera but clung to film. Nokia dominated mobile phones but missed the smartphone revolution. Yahoo! had the chance to buy Google – twice.

But here's what makes the AI revolution different, and frankly, more dangerous. Those earlier technological shifts happened over years, sometimes decades. Companies had time to adapt, even if many chose not to. The AI revolution? It's happening in months, sometimes weeks.

Think about this. ChatGPT reached one million users in just five days. It took Netflix three and a half years to hit that milestone. The pace of AI advancement isn't just fast – it's exponential. What was cutting edge six months ago is now obsolete. The tools and capabilities that seem impressive today will be basic expectations tomorrow.

And just like Blockbuster in 2000, I'm seeing businesses today dismiss AI as a "future concern" or a passing trend. They're making the same mistake: assuming they have time to adapt later. But here's the hard truth. In the AI era, later often means too late.

This chapter isn't about scaring you – it's about preparing you. Because unlike Blockbuster, you have a choice. You can see the change coming and position yourself ahead of it. The question isn't whether AI will transform your business … it's whether you'll be the Netflix or the Blockbuster of your industry.

Understanding AI's impact

AI is a fundamental shift in how work gets done. For the first time in business history, we're seeing something remarkable: the ability to do things faster, cheaper and better. Typically, you could only pick two of these. Want it faster and cheaper? Quality would suffer. Want it better and faster? That would cost you. But AI is rewriting these rules.

To understand why AI represents such a dramatic shift in business capabilities, we need to look at what came before. Some businesses are familiar with Robotic Process Automation (RPA) – software robots that follow precise instructions to automate repetitive tasks. Think of RPA like a very diligent worker who follows your documented process exactly, without deviation. It is fantastic for tasks like data entry, file manipulation or moving information between systems.

But here is the limitation: RPA is completely inflexible. If something in the process changes, even something as small as a form field moving position or a button being added, the automation breaks. It is like a train that can only run on its predetermined tracks. The moment it encounters something unexpected, everything stops.

AI, on the other hand, is more like an adaptive problem solver. It can take in all available data, recognise patterns and make decisions, even in situations it hasn't explicitly been programmed for. When it encounters something unexpected, instead of breaking, it can analyse the situation and determine the best way forward.

Let me give you a practical example. An RPA bot processing customer refund requests would follow rigid steps: accessing the CRM system, locating the customer record, verifying purchase history, calculating the refund amount based on policy and submitting the approval form. If the CRM interface changes even slightly, like the "Process Refund" button moving from the bottom to the side panel, the bot completely fails. But an AI system handling refunds can adapt by analysing screen elements to find the relocated button. AI can understand customer sentiment in their request email, it can determine refund eligibility even for unusual cases (like purchases just outside the policy window) and suggest appropriate goodwill gestures based on customer history and value … all without being explicitly programmed for each scenario.

Beyond these operational tasks, AI truly shines in administrative and writing tasks, areas that have historically resisted automation because they require understanding context and making judgement calls. While RPA might help you move data between spreadsheets, AI can read reports, draft responses and make informed decisions. It can handle nuanced tasks like crafting customer service replies, generating marketing content or summarising lengthy documents – tasks that previously required significant human time and effort.

This distinction helps explain why AI is transforming every corner of business. It's not just faster automation: it's a technology that can understand context, learn from experience and make informed decisions.

Here's an example of how we transformed our own content creation process here at SYSTEMology. Like many businesses, we were creating content to generate leads, videos, blog posts, social media updates and email newsletters. It was a complex, time-consuming process involving multiple steps like video editing, transcribing, writing titles and descriptions, creating social media posts and crafting email marketing campaigns. (Learn more about the process in my book *Authority Content*.)

At its core, the system was strong. But as we started exploring AI tools, we realised we could re-engineer the entire process. We began by reviewing our existing way of doing things and systematically identified where AI could help. After we trained the AI on our data, created prompts and updated our workflows, the results were remarkable.

Tasks that once took hours now take minutes. Best of all, the new process is not just faster and cheaper – it's better. Because our team now spends less time on routine tasks like transcription and initial drafts, they can focus more energy on refinement and creativity. The AI handles the heavy lifting, while humans add the strategic thinking and personal touch that makes content truly exceptional. And that is just with one of our processes.

That's just one example of the change that is happening. This transformation is happening in every corner of business. Marketing teams are using AI to create personalised campaigns at scale. Sales teams are leveraging AI to qualify leads and predict customer behaviour. Operations teams are automating routine tasks and detecting inefficiencies. Finance departments are using AI to spot patterns in data and forecast trends. HR teams are streamlining recruitment and improving team member engagement. The impacts are profound.

I like to think that AI isn't replacing humans but augmenting them. It's like giving everyone on your team a super-powered assistant. These assistants can handle routine tasks, provide insights and help make better decisions.

AI as a data refinery

AI is like a modern-day oil refinery. Just as a refinery takes crude oil and transforms it into valuable products such as gasoline, plastics and chemicals, AI takes raw data and refines it into valuable insights and actions.

Now think about all the data sitting in your business right now, including:

- Customer interactions and feedback
- Sales patterns and trends
- Financial transactions and reports
- Team performance metrics
- Marketing campaign results.

Without AI, much of this data sits untouched, its potential value locked away. It's like having an oil field but no refinery. But with AI, you can instantly transform this raw data into actionable insights. Patterns emerge. Trends become visible. Opportunities reveal themselves.

And that's just the tip of the iceberg. The most exciting part, and what makes your role as Systems Champion so crucial, is that AI needs clear instructions to tell it what to do. It needs documented processes. Every Standard Operating Procedure (SOP), every documented workflow and every captured process becomes training material for AI. You're not just creating instructions for humans anymore; you're creating the programming that will power your AI assistants.

Think about that for a moment. When you document how your best customer service representative handles difficult conversations, you're not just creating a training manual – you're creating the foundation for an AI that can handle similar conversations. When you document your sales team's follow-up process, you're laying the groundwork for automated lead nurturing. When you combine this training material with actual client data, that's where the magic happens.

Can you see it? This is why clear, well-documented systems are more valuable than ever.

Your unique position

Let me share something that took me years to fully understand. The most valuable person in a business isn't always the one with the most technical skills or industry experience. It's often the one who understands how everything fits together. And that is exactly where you sit as a Systems Champion.

Think about what you do every day. You're not just documenting processes – you're mapping the DNA of your business. You understand how work flows from department to department. You see the connections that others miss. You know which processes are working smoothly and which ones need improvement. This unique perspective puts you in the perfect position to lead your business's AI transformation.

Here is why. To be effective, AI needs:

1. Clear instructions about what to do
2. Understanding of how different parts connect
3. High-quality data to learn from and work with.

You have access to all this at your fingertips.

Remember the case study I shared earlier with Eryn from Stannard Homes? She started in the business doing interior design but was curious about everything. By documenting systems across different departments, she gained such a comprehensive understanding of the business that she became one of its most knowledgeable team members.

Ryan (the business owner) is mentoring her to manage their $15–20 million operation while he launches a new venture. And whether she realises it or not, her deep understanding of how everything connects makes her ideally positioned to spot where AI can add the most value.

This is the big opportunity in front of you. This transition toward an AI-driven workplace will necessitate the rise of a new generation of champions – individuals who possess a profound understanding of both the complexities of business processes and the vast potential of AI.

Think of yourself as a bridge between the current way of working and the AI-powered future. You're not just a Systems Champion anymore; you're becoming a Systems and AI Champion. This isn't just about job security, but about career opportunities. The skills you're developing now, combined with your growing understanding of AI, position you to lead your business's transformation. You could become one of the most crucial players in your organisation's future success.

Your starting point

When I talk about AI transformation, I often see a mix of excitement and overwhelm in people's eyes. They know this is important, but where do they begin? Here's the good news. As a Systems Champion, you've already started.

I've already introduced the concept of using AI tools like ChatGPT to help with systems documentation. In the past, turning video recordings into step-by-step processes took hours or even days of careful study and manual transcription, just to get a first documentation draft for a single system. Now, with AI assistance, you can upload a recording, get an accurate transcript and have the AI organise it into clear procedural steps, getting you 80–90 percent of the way there in a matter of minutes.

This is a prime example of AI at work.

You have probably also seen how many AI capabilities are being added to your existing tools. Software like Microsoft Office, Google

Workspace and many project management platforms have built-in AI features. Start experimenting with these. They are a low-risk way to begin understanding AI's potential.

For example, that transcription feature in Microsoft Teams? That's AI. The smart compose in Gmail? Also AI. These might seem small, but they're perfect training grounds for understanding how AI can augment human work.

While you're documenting systems, start organising them with AI in mind. Think of it like creating a knowledge base that both humans and AI can learn from. The more organised and structured your documentation, the easier it will be to use as AI training material later.

Look to add AI tools into existing processes to create a quick win. Maybe it's using AI to help draft customer service replies, or to summarise meeting notes, or to help with data entry. Success in these small areas builds confidence and creates momentum for bigger projects.

Your goal isn't to become an AI expert overnight. It's to gradually build your understanding while leveraging your unique position as Systems Champion.

Creating AI-powered team members

Imagine creating a digital twin of your best team members – one that is available 24/7 and can handle multiple enquiries simultaneously. That's what happens when you create AI assistants that are specifically trained on your business's knowledge and processes.

Using your systems management software, you can organise your documented processes by role or department, then train AI specifically on that knowledge. Want an AI assistant that thinks like your best customer service representative? Feed it your customer service SOPs, call recordings and common scenarios. Need an AI that understands your sales process? Train it on your sales playbooks, client personas and pricing documents.

The key is organisation, and you've already started building a database of processes for different departments and roles. Think about having different AI assistants, each an expert in their domain. These are powerful assistants that help your team work more effectively. They know exactly how your business works because they've been trained on your specific systems and processes.

Pretty cool, huh?! And this is just the start!

Build an AI-driven culture

You'll quickly learn that the true power of AI emerges when it becomes woven into your organisation's cultural fabric. There's a whole host of opportunities here. You just need your entire team embracing these new capabilities and thinking differently about how work gets done.

I'll share more about this a little later. As you know, "Culture" is one of the three key pillars covered in this book, and I've dedicated an entire section to this topic. That said, while we're talking about AI, I wanted to share six practical tips for approaching AI from a cultural perspective that you can implement immediately.

Tip #1 – Start with quick wins: Begin with tasks that everyone agrees are tedious or time-consuming. When people see AI eliminating their least favourite parts of work, resistance naturally drops. Show them how AI can draft first versions of routine emails or summarise long meetings. These small wins build confidence and curiosity about what else is possible.

Tip #2 – Make it personal: Help each team member understand how AI will make their specific role better. Show your salespeople how AI can help them prepare for client meetings. Show your customer service team how AI can help them respond to enquiries faster. When people see direct benefits to their daily work, they become supporters rather than resisters.

Tip #3 – Address fears head-on: Yes, people worry about AI replacing their jobs. Don't dodge these concerns. Instead, address them directly. Show how AI is about augmentation, not replacement. Share examples of how team members who embrace AI are becoming more valuable, not less. Like Eryn at Stannard Homes, they often find themselves moving into more strategic roles.

Tip #4 – Create safe spaces to learn: Give your team permission to experiment and make mistakes with AI. Create informal learning sessions where people can share discoveries and ask questions. The goal is to make AI feel like a helpful tool, not an intimidating technology.

Tip #5 – Celebrate and share successes: When someone finds a clever way to use AI, make it known. Create communication channels for sharing AI wins and insights. This not only spreads knowledge but also builds momentum for adoption.

Tip #6 – Keep the human touch: Always emphasise that AI is a tool to enhance human capabilities, not replace them. Show how AI handles routine tasks so people can focus on what humans do best: building relationships, solving complex problems and being creative.

Your role as Systems Champion puts you in the perfect position to lead this cultural shift. Embrace this opportunity. Think seriously about expanding your ambitions from Systems Champion to AI Champion as well. The future of your business, and indeed the future of work itself, is being written by Systems Champions like you.

AI Champion Action Plan

Pick one task where AI could help immediately:

....

....

....

Which team members would benefit most?

....

....

....

What would be the measurable positive impacts?

❑   Time saved

❑   Quality improved

❑   Tasks automated

❑   Team engagement increased

❑   Other: ....

Not sure where to start? Go to ChatGPT or your favourite AI and ask:

"I work in a <insert your business industry> business and I would like to know the top three areas where I could use AI to see immediate positive results. Please also let me know which team members you think would benefit and how I could measure the positive effects of those changes."

16

Workshops & Training

Y OU KNOW THAT SAYING ABOUT it taking 21 days to form a habit? The one that gets quoted in every self-help book and motivational speech? Well, it's not quite right. It originated with Dr Maxwell Maltz, a plastic surgeon who published a book called *Psycho-Cybernetics* in 1960. Dr Maltz observed that his patients took "about 21 days" to adjust to seeing their new faces after surgery, writing that "it usually requires a minimum of about 21 days for an old mental image to dissolve and a new one to jell." Over time, this was simplified into the now-ubiquitous claim that "it takes 21 days to form a habit."

But it's not really supported by more modern research. A 2009 study from University College London tracked 96 people forming new habits and found it actually takes closer to 66 days for a behaviour to become automatic. For some people, it was as fast as 18 days. For others, it took 254 days. While not a massive sample size, this peer-reviewed research provides a more evidence-based foundation than the arbitrary 21-day myth.

I share this because, as you know, your role as Systems Champion isn't just about documenting processes. It's about helping people form new habits. You're asking team members to change how they work, to adopt new ways of doing things. And while we don't need to wait 66 days to see results, we do need to support this change over a longer period of time.

This reality of habit formation highlights why one-off introductions to new systems rarely stick. Sending an email with links to new documentation or holding a single meeting to announce changes simply isn't enough to rewire established work patterns. Systems adoption requires consistent reinforcement, practice opportunities and support throughout that critical 66-day window when new behaviours are still fragile and easily abandoned.

I remember working with a Systems Champion who had created beautifully detailed documentation for their customer service processes. The processes were perfect, the workflows were clear … but three months later, team members were still doing things their own way. Why? Because the Systems Champion never created a structured way to bring everyone along on the journey.

This is where workshops become your secret weapon. They give you a platform to announce new initiatives, engage your team and dramatically increase the likelihood of successful implementation. Whether you're working with entire departments or smaller groups, these focused sessions create momentum for your systemisation efforts.

What exactly is a workshop?

Now, I get it … for some, the idea of running a workshop is a stretch outside your comfort zone. But let me put your mind at ease. A workshop can be as simple as a focused conversation about systems.

Sometimes you need a structured training session. Other times, a casual Q&A works better. You might use group discussions to solve specific challenges, quick demonstrations to show a new process or interactive exercises to practise using tools. The format should match what your team needs in that moment.

For your first one, make it easy on yourself and begin by asking for 15 minutes in your next team meeting to share a system win, or join a department huddle to demonstrate a new tool. Remember, your goal isn't to become a professional trainer. It's simply to get people thinking and talking about systems.

I remember a Systems Champion who started just by sharing a five-minute systems tip at the end of each weekly meeting. Those brief moments created more engagement than her previous attempts at formal training sessions. Why? Because she made systems approachable and relevant to everyone's daily work.

I can't understate how powerful workshops can be. There's a well-known saying: "The only thing worse than training your employees and having them leave is not training them and having them stay." And it's true: the cost of mistakes, inefficiencies and double work far exceeds what it takes to train your team correctly.

Planning your workshop with AI

So where do you start? Here are some questions to consider:

- What specific outcome do you want?
- Who needs to be there?
- What's the minimum time needed?
- What practical examples can you use?
- How can you make it interactive?

Just start brainstorming some ideas and putting together a rough plan.

The first workshop you run could be about SYSTEMology and your Minimum Viable Systems (MVS). The second could be showcasing a system or two. The third could be teaching some of the tools you're introducing. The fourth could be about some of your findings about AI.

Preparing a workshop outline is much easier than you think with a little help from AI. I gave the AI a little context about what I was working on, shared some resources and then asked …

"I'm a Systems Champion looking to introduce systemisation to my team. Can you create a detailed outline for a 30-minute workshop that's simple and engaging? It should focus on explaining the benefits of systemisation, involving the team in identifying one process to systemise and setting clear next steps. Keep the language approachable and practical, with a balance of talking points and interactive activities. The goal is to inspire action and make systemisation feel achievable."

Here's what I got: a streamlined 30-minute workshop outline.

Workshop Title:

"Introduction to Systemisation: Small Steps, Big Impact"

Duration: 30 minutes

Objective:

- Briefly introduce systemisation and its benefits.
- Engage the team with a simple group activity.
- Identify one process to systemise as a starting point.

Agenda Outline:

1. Welcome & Purpose (5 mins)

Purpose: Set the tone and explain why everyone's here.

Steps:

1. Greet the team with energy.

2. Share the purpose: "We're here to start making our work easier and more efficient through systemisation."

3. Ask a quick icebreaker: "What's one repetitive task you wish could handle itself?" (Ask for 2–3 volunteers to share.)

2. The Power of Systems (5 mins)

Purpose: Highlight the benefits in simple terms.

Key Talking Points:

- Systems reduce stress and mistakes, save time and create clarity.

- Share an example of a system's impact (e.g. using a checklist to onboard a client more efficiently).

- "Systemisation doesn't mean more work – it's about making your life easier."

3. Quick Group Activity: Pick a Starting Point (15 mins)

Purpose: Get the team involved and identify one process to systemise.

Steps:

1. Brainstorm (5 mins):

- Ask: "What's one process or task in your role that could benefit from having clear steps?"

- Write suggestions on a whiteboard or shared document (e.g. handling enquiries, setting up a new project or managing client feedback).

2. Vote (2 mins):

- Quickly vote on the most impactful or easiest process to document.

3. Map It Out (8 mins):

- Work together to outline the high-level steps of the chosen process.
- Example: "Customer Enquiry Handling: Step 1 – Answer the call; Step 2 – Collect details; Step 3 – Log in CRM; Step 4 – Follow-up email."

4. Commit to Action (5 mins)

Purpose: Create momentum and accountability.

Steps:

1. Assign roles: Who will document the steps, who will test the process and when it will be reviewed.

2. Share next steps: "In our next meeting, we'll review the draft of this system together."

3. Encourage participation: "This is the first step toward making everyone's job easier!"

Follow-Up Actions:

1. Email a summary of the chosen process and action items.

2. Set a date for a follow-up workshop to review progress (ideally 1–2 weeks later).

Now it's your turn

Pretty cool, huh? And what a fantastic starting point. You can find more workshop templates and resources here: **www.SystemsChampion.com/ resources**

Just remember to keep your workshops easy, fun and engaging. For better or worse, many people have baggage around the topic of systems and processes, and we need to challenge those false beliefs. We want to show them the SYSTEMology way of doing things!

And once you find your groove, running a workshop weekly would be a great target for you to set for yourself. Research shows that smaller, regular training sessions are far more effective for building and retaining new habits than infrequent, intensive ones. This approach, known as spaced repetition, not only improves long-term retention but also makes it easier for your team to integrate changes into their daily routines.

Your First Workshop Blueprint

Topic: ....

Time needed: ....

Key participants:

....

....

....

Format:

❑   Team meeting segment

❑   Dedicated session

❑   Department huddle

❑   Other: ....

Main points to cover:

1 ....

....

2 ....

....

3 ....

....

Interactive elements:

❑   Group discussion

❑   Live demonstration

❑   Hands-on practice

❑   Q&A session

❑   Other: ....

Case Study From Systems to Automation

When Shannon Smit founded Smart Business Solutions, she was a solo accountant with a vision of building a different kind of accounting practice. Today, her Melbourne-based firm serves thousands of clients, employs over 20 team members and is leading the way in business automation. But the path to this success required embracing both systems and technology in a way that would transform how they operated.

Breaking free from manual tasks

Like many accounting practices, Smart Business Solutions faced the constant challenge of balancing high-value client work with necessary but time-consuming administrative tasks. Despite having a structured approach (Shannon credits her early McDonald's experience for this), the business struggled with the sheer volume of repetitive work. Team members were spending hundreds of hours annually on mundane tasks that, while essential, took them away from meaningful client interactions.

"We had great people doing low-value work," Shannon recalls. "With the shortage of accountants and increasing wage pressures, we knew something had to change."

Enter the systems foundation

Shannon discovered systemHUB and began the crucial first step: documenting their processes. Over six to seven years, the team built such a comprehensive library of systems that "Is it in systemHUB?" became their catchphrase. This documentation proved transformative, making the invisible visible and revealing opportunities for improvement.

With clear processes documented, Smart Business Solutions began automating processes. At first, team members were hesitant, fearing job losses. But Shannon's vision wasn't about replacing people: it was about elevating their work.

The results were dramatic. One automated process alone saved 998 hours annually, equivalent to nearly 60 percent of a full-time team member's working hours. Another routine administrative task that consumed two hours daily was completely automated, saving 10 hours per week. The business even automated an entire administrative role, allowing team members to focus on higher-value work.

Cultural transformation

What started as fear of automation transformed into enthusiasm. The team now actively suggests processes for automation, particularly tasks they find mundane or repetitive. "If you find it boring because you're taking numbers from one spot and putting them into another spot, that's what our robot should be doing," Shannon explains. "You should be spending time talking with clients and using your brain on more exciting work."

The firm's systematic approach to identifying automation opportunities includes:

- Measuring hours spent on each process
- Identifying low-value, repetitive tasks
- Prioritising based on ROI
- Using existing system documentation to guide automation.

Looking toward the future

Smart Business Solutions continues to evolve, now exploring AI tools like Microsoft Copilot while maintaining their commitment to personal service. The team that once feared automation now sees it as a path to more meaningful work. "People still need people," Shannon notes. "We can offer cost-effective service because we've automated the routine tasks, allowing our team to focus on what truly matters – client relationships."

Through the combination of systematic thinking and strategic automation, Smart Business Solutions has transformed from a traditional accounting practice into a technology-enabled firm ready for the future of work.

Watch the full interview here:

www.SystemsChampion.com/resources

Pillar 2 Action Items

☐ Audit your current tools against the core requirements: systems management, project management, recording and office suite capabilities. Identify any critical gaps.

☐ Create your AI implementation roadmap, starting with quick wins. Begin with tasks that link to your tasks as a Systems Champion.

☐ Design your first systems workshop. Use AI to help create an engaging 30-minute session introducing your systemisation journey to the team.

☐ Establish your training rhythm. Schedule regular, focused sessions to build momentum and support lasting habit formation.

☐ Build a feedback loop with your team. Set up regular check-ins to understand how new tools and AI implementations are being adopted.

☐ Start building your AI-driven culture. Share quick wins, address concerns openly and create safe spaces for learning and experimentation.

PILLAR 3: CULTURE

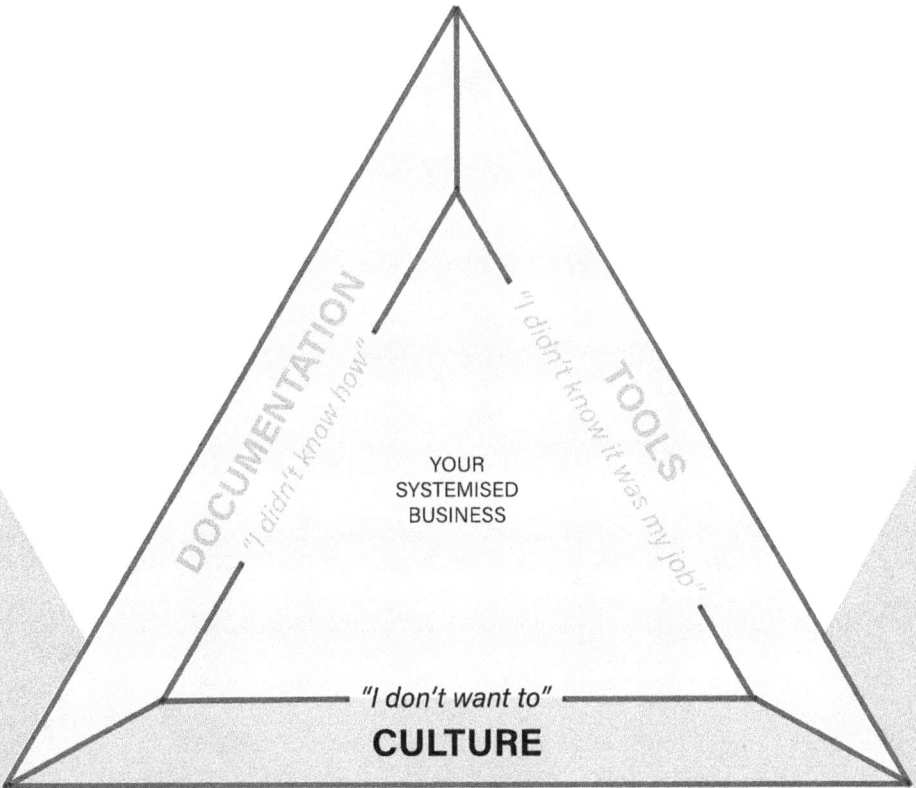

DOCUMENTATION

"I didn't know how"

TOOLS

"I didn't know it was my job"

YOUR
SYSTEMISED
BUSINESS

"I don't want to"

CULTURE

"Example is not the main thing in influencing others. It is the only thing."

Albert Schweitzer, Nobel Peace Prize laureate

Summary

Building a strong systems culture requires more than just processes. It demands deliberate attention to people, culture and accountability. Success comes from creating an environment where systems become "the way we do things here".

Highlights covered in these chapters include:

- The "dancing guy" principle: how cultural movements start and gain momentum.

- Why conviction and deep belief are essential for Systems Champions.

- The three steps to cultural transformation: committing to your beliefs, finding supporters and building proof.

- How recruitment and onboarding can reinforce systems culture.

- The System for Unfollowed Systems (SFUS) framework.

- How to handle persistent resistance while maintaining team morale.

- The critical balance between support and accountability in systems adoption.

17

Power of Culture

———————

OKAY, YOU'VE REACHED THE FINAL boss!

You've created clear documentation and placed it exactly where your team needs it, eliminating the "I didn't know how" excuse. You've implemented the right tools for transparency and accountability, removing the "I didn't know it was my job" objection.

But now the biggest challenge of them all … "I don't want to."

How do we get people to want to follow systems?

First things first, I'm going to get you to watch something. Head over to: **www.SystemsChampion.com/dancingman** and spend three minutes watching what might be one of the most powerful illustrations of how to build a movement. I'll wait.

…

…

…

I'm only going to continue once you've watched it. Watch it now and then come back. I'll just wait for you.

…

…

…

You didn't actually watch it yet, did you? That's fine, I've got all day …

…

…

…

Okay, now let's break down what you just watched.

It's a perfect summer afternoon at the 2009 Sasquatch! Music Festival. The Gorge Amphitheatre stretches out like a natural colosseum, its rolling hills dotted with hundreds of festivalgoers lounging on blankets. The spectacular Columbia River winds below, creating a backdrop that makes even the most cynical attendee pause in wonder.

In this setting, something extraordinary is about to unfold.

A lone figure rises from the crowd. He's shirtless, confident and about to challenge every social norm on that hillside. As Santigold's "Unstoppable" drifts from the main stage, the man begins to dance. His movements are a curious blend of tribal rhythm and modern freestyle: unconventional, authentic and utterly uninhibited. It takes guts to stand alone and look ridiculous, but what he's doing is simple, almost instructional – a crucial quality when you want others to follow.

The crowd's reaction is predictable. Some point and laugh. Others avert their eyes, that uniquely urban response that says, "I see you, but I'm pretending not to." For a short while, he remains alone in his dance, like a solitary tree swaying in still air, defying the comfortable inertia of the crowd.

But this is when the magic begins.

A second young man, wearing a green T-shirt that would become iconic in leadership studies, makes what business schools now analyse as a perfect example of early adoption. He joins the dance. His action, seemingly simple, is revolutionary. In that moment, he transforms what everyone sees. The "crazy dancing guy" becomes the leader of something. The first follower transforms a lone nut into a leader.

The shirtless dancer's response is crucial. He doesn't try to lead or direct. Instead, he welcomes his first follower as an equal partner. Together, they create something new – not a performance but a shared experience. It's no longer about the leader. It's about them, plural. What was once perceived as one person's questionable choice becomes a deliberate movement.

Within the first minute, the third person joins – a critical turning point. Now we're witnessing a phenomenon that behavioural scientists study intensely. When people see others doing something, they become significantly more likely to join in themselves. It's human nature. Two more join, and then almost immediately after, another group of three.

Each new participant lowers the psychological barrier for others, transforming scepticism into curiosity, and curiosity into participation. Now new followers copy the other followers, not the leader.

By the one-and-a-half-minute mark, people aren't just joining, they're running to join. What marketers call FOMO (Fear of Missing Out) becomes visible energy. The movement hits its tipping point – that magical moment when social friction gives way to unstoppable momentum. Joining is no longer risky: it's the safe choice.

In the final minute, the transformation completes itself. The hillside that had been a passive audience becomes a pulsing mass of hundreds of dancers. The original "crazy dancing guy" is no longer visible. The movement has transcended its founder, becoming something bigger than any individual.

Your business owner is that first dancer

Here's why that story matters to you right now. Your business owner has taken the bold step of standing on the figurative hillside and declaring, "We're becoming a systems-first business." Like that shirtless dancer, they're challenging the status quo, standing up against the comfortable inertia of "That's how we've always done things."

And you? You're about to play an even more crucial role. As the Systems Champion, you're that vital first follower. The person who transforms a solo act into a movement. Your role isn't just about implementing systems. It's about showing others that this new way of working isn't just safe: it's revolutionary!

Think about what that first festival follower did. He didn't just dance: he demonstrated. He showed everyone else exactly how to join in. That's precisely your role with systems implementation. You'll show your team how to embrace this new way of working, make it accessible and celebrate those who join the movement.

This is how you'll build your systems culture. Not through mandate, but through momentum. Starting with you, the crucial first follower, you'll systematically create an environment where "This is how we do things here" becomes your organisation's natural rhythm.

Sounds easy enough, right? Just follow the shirtless guy.

But the reality is that changing culture requires more than enthusiasm. It requires strategy. Building a systems culture addresses the core "I don't want to" resistance you'll encounter. People are motivated more powerfully by social belonging than by almost any other force. When you successfully transform "how we do things here", you create an environment where following systems becomes the path of least resistance rather than the exception.

Are you ready to join the dance?

Let me share with you the three steps to start a cultural movement, starting with what might surprise you as the most crucial element … you.

Step 1: Commit to your beliefs

Here's something unexpected I've discovered when working with Systems Champions. Technical skill isn't the primary predictor of success. Neither is experience. The most successful Systems Champions share something far more fundamental: they deeply believe in what they're doing.

Remember our dancing guy story? That first follower didn't join because he had studied contemporary dance or mastered rhythm. He joined because he believed in the joy of what was happening. Your journey as a Systems Champion begins in exactly the same place: with belief.

But wait … isn't this a given? You're the Systems Champion. Of course you're going to be a wholehearted believer. The proof is right there in your job title.

It isn't that simple.

When you landed your new role and agreed to be paid for your systemisation work, you committed to the process. But it doesn't automatically follow that you believe in it and can actively defend it. How deeply have you really thought about what you're doing and whether it's really the right way forward for the business? We've discussed some of the arguments for systemisation and it may all sound reasonable and logical, but that isn't enough.

It's a bit like taking a political position on an issue because it looks and feels right, but without really understanding the underlying facts. When this happens, it's very easy for someone more knowledgeable to come along and dismantle your belief system and your faith in your stance.

You don't really believe in something unless you can defend it. And you can't defend something with any conviction unless you properly comprehend it.

This is going to be the focus of this step. You need to be a champion of systems, not just because it's your job, but because you understand, on a deep level, why it's so important. You need to be able to defend it against every objection that anyone can throw at you.

Culture, just like the other two pillars of SYSTEMology, is a crucial piece of the puzzle and is every bit as important. It's also perhaps the most challenging. The culture that exists within a workplace is able to trump just about everything else that a business does to try to change things. You can run all the workshops and initiatives that you like to try to drive new behaviours, but if the underlying culture doesn't change, then everything will keep snapping back into the old way of doing things.

Virtually every Systems Champion runs into this kind of resistance from their teams, not just at the outset of the project, but weeks or even months into the process. The bigger the team, the more likely this is to occur and the harder it will be to overcome. It's very hard to stay the course and get over the initial hump unless you're really confident in what you're preaching. If you're going to change the culture in your business to one that believes in and fully embraces systemisation, you must first believe in it yourself.

Let me share with you the core beliefs I've seen in the most remarkable Systems Champions – the beliefs that drive true transformation.

Core beliefs of remarkable Systems Champions

Systems are building blocks: Every successful and profitable business is built on systems and processes. They're not optional extras. They're the fundamental elements that help you find and keep customers, eliminate waste and dominate your market. When you truly understand this, you see that systems aren't just about organisation, but about creating repeatable success that sets you apart from your competition.

Every problem is a systems problem: This belief changes everything about how you handle challenges. When something goes wrong, you don't look for someone to blame: you look at the system. Late deliveries are not about lazy team members – they're about scheduling systems that need improvement. Quality issues are not about careless workers – they are about processes that need refinement. This perspective transforms finger-pointing into collaborative problem-solving.

Systems make your work easier: Systems aren't about adding complexity. They're about radical simplification. The best Systems Champions understand this paradox. Structure creates freedom. When a system is working well, work flows effortlessly. When everyone knows exactly how things work, they can focus their energy on innovation and improvement rather than figuring out what to do next.

Systems make you more valuable: Systemising a role doesn't make you replaceable; it makes you irreplaceable. By creating clear, transferrable systems, you demonstrate your ability to think strategically and elevate entire processes. You're no longer just a doer of work but an architect of efficiency. You're showing the organisation how work can be done better, faster and more effectively. This is your path to becoming indispensable.

Systems development is ongoing: Systems aren't a one-and-done project. They represent a fundamental shift in how we think about work. True transformation happens through consistent, incremental improvement. The most effective organisations don't chase perfection – they pursue evolution. Small improvements, made consistently, compound over time to create breakthrough results.

Every business is a school: This final belief will transform how you see your role. When you're building systems, you're creating learning opportunities. As Michael E. Gerber taught me, businesses aren't just commercial enterprises – they're learning platforms. Every system you build is a lesson plan that teaches new team members how to add value to your organisation. Every life a legacy, every business a school.

Now, as I said at the beginning of this chapter, being granted the role of Systems Champion does not automatically make you a fully-fledged supporter of systemisation in business. And, sorry to say, reading and acknowledging these six core beliefs isn't going to do it either.

But it's a move in the right direction.

There's something of a leap of faith required here since they may or may not fit in with your existing beliefs. And you won't truly be able to say that you fully buy into them until you've acted on them and seen the results for yourself.

Take, for instance, the second core belief: Every problem is a systems problem. You may or may not find this belief easy to accept. But either way, you won't be able to firmly internalise it until you've tested it in the real world and seen the results. When a problem occurs, you'll need to tackle it from the perspective that the issue is a system issue, even if you instinctively feel like the error is down to a human failing.

This isn't quite a "fake it till you make it" scenario, because you're starting from a position of knowledge and acceptance. But to truly become a Systems Champion, in more than just job title, you're going to have at least act as if you believe these tenets to be trustworthy.

Then, once you see the results, you'll build confidence in these core beliefs based on your own experience, not just on what you've read in these pages.

And when you truly believe, something remarkable happens. It shows up in ways you can't even articulate. Your body language shifts. Your tone changes. You respond to challenges with quiet confidence. Under pressure, you remain unshakeable. Your actions become a seamless extension of your beliefs.

Ask yourself: How would I show up differently if these system beliefs were my absolute truth? What would change in my approach to work, to challenges, to growth? This is where your journey as a Systems Champion truly begins. Not with tools or techniques, but with these core beliefs. Let them be your foundation as we move forward into the practical work of transformation.

Let's assess where you truly stand with these core beliefs.

Your Systems Champion Conviction Check

Rate your current level of conviction from 1–10:

Sceptical → Understanding → True Believer

Systems are building blocks: ....

Every problem is a systems problem: ....

Systems make your work easier: ....

Systems make you more valuable: ....

Systems development is ongoing: ....

Every business is a school: ....

Which belief resonates most strongly with you today?

....

....

....

Which one challenges your current thinking the most?

....

....

....

What personal experience has validated one of these beliefs?

....

....

....

Step 2: Find your supporters

Okay, back to our dancing guy. There's something crucial to remember. That shirtless dancer didn't win over the entire hillside at once. He started with just one person, then another, then another. Each new dancer made it easier for the next person to join.

This is exactly how you're going to build your systems culture. And here's the hard truth: not everyone will be excited about systems right away. And that's okay. Your job isn't to convince the sceptics or argue with the resisters. Your job is to find your fellow dancers.

Who in your organisation naturally gravitates toward order and improvement? Who lights up when you talk about making work easier and more efficient? Who's already shown interest in better ways of doing things?

These are your people. These are your first followers.

Start with them. Work closely with them. Show them exactly how to succeed with systems. Make it easy, make it obvious and make it fun! Remember, you're not just implementing processes. You're starting a movement. And movements thrive on enthusiasm, not enforcement.

Here's how to make it work.

Make it easy: Break down system adoption into small, achievable steps. Don't overwhelm your supporters with complexity. Give them simple wins they can achieve quickly. Each small success builds confidence and creates momentum. The key is to reduce friction at every turn. If adopting a system takes more than two steps, simplify it. Your goal is to make following the system the path of least resistance.

Make it obvious: Design your workspaces to trigger system use. Create visual cues that prompt the right behaviours. Place system documentation exactly where it's needed. Put checklists directly in workspaces. Ensure team members are never more than one click away. In fact, make system use so obvious that it would feel strange not to do it.

Make it fun: Yes, fun. Systems don't have to be dry and boring. Celebrate wins, showcase improvements and make progress visible. Create instant positive feedback. Find ways to make the benefits of using systems immediately satisfying. Turn system adoption into a positive experience that people want to be part of. Remember, we repeat behaviours that make us feel good.

And here's my best advice: don't waste energy trying to convince the resisters. Not yet anyway. It's a common mistake to focus on the people pushing back, but that's not how movements start. Movements start with the willing.

As you work with your supporters and celebrate their successes, that's where the magic happens. Other team members start noticing. They see their colleagues creating new systems, solving problems and getting praised for it. They see the benefits of systems playing out in real time.

This is the showcase effect, and it's more powerful than any argument you could make. When people see their peers succeeding with systems, their scepticism naturally begins to fade. The question changes from "Why should we do this?" to "How can I join in?"

And those initial resisters? Many will come around on their own as they see the movement growing. Those who don't will find themselves increasingly out of step with the new normal. Either way, you won't need to spend energy convincing them – the culture will do that work for you.

Movements aren't built by converting critics. They're built by nurturing supporters. Find yours, work with them, celebrate them.

Step 3: Build your proof

You've arrived at the final step of your cultural transformation process. And I want you to imagine something with me. Picture yourself in a courtroom. The judge looks down from the bench and asks, "Can you prove, beyond a shadow of doubt, that your company is truly systems-driven?"

Take a moment to consider what evidence you already have. By reading this book and embarking on this journey, you've already started building your case. You have:

- Been selected as a Systems Champion

- Documented some key processes

- Started installing some new tools

- Run a couple of workshops

- Started engaging your first followers

- Celebrated some early wins.

That's a solid start. But as any good lawyer will tell you, winning cases requires overwhelming evidence. So what else could you do? What additional proof could you keep building? Think about how you could:

- Share system wins in your company chat channels

- Make systems part of every team meeting agenda

- Create a company systems newsletter

- Document your systems journey with before-and-after results

- Showcase great examples of systemisation

- Give everyone a copy of this book to read

- Host a team celebration as you hit key milestones

- Publicly track your progress and turn systemisation into a game

- Give prizes and rewards for new system initiatives

- Create a clear process for when systems aren't followed

- Get T-shirts, notepads and other stationery that remind people to follow process

- Include system metrics in team members' performance reviews

- Build system mastery into your career advancement paths

- Add systems-thinking as a company value

- Project a systems-driven identity externally to clients, suppliers and partners.

Your job as a Systems Champion is to keep building this proof. To keep planting evidence that systems aren't just something you do; they're fundamental to who you are as an organisation.

And now for the million-dollar question. How long should you keep building proof?

Unfortunately, there is no precise timeline or specific identification that you've "made it" since it can vary largely depending on the size, make-up and acceptance by your existing team. You might be able to point to key metrics that have improved since you started your project and draw a straight line between your work and the results, but that isn't enough to say that the business has become systems-focused down to the cultural level.

Remember, making SYSTEMology part of the business culture is the only way to ensure that all the hard work you're putting in stays around for the long term. It needs to become so ingrained that whenever anyone in the company is tackling a new project or aiming to fix a new problem, their first thought is how they can develop or adjust a system to reach their goal.

A new culture's main enemy is always the existing culture. There's a stultifying phrase often thought, but rarely said out loud, which makes integrating systems so challenging: "That's just how we do things around here." Paradoxically, this is exactly the thought-terminating cliche that you want to embed into the new culture you're developing. Your goal is to eventually reach a point where everyone within the business instinctively uses and develops systems as part of the work because they know … "This is how we do things here."

That's what a business with a culture of systemisation looks like. It isn't some magical switch that takes a business from chaos to a well-oiled machine. It's a gradual development that will reveal itself by how people use systems in their day-to-day work without even questioning whether systems are the answer.

You can't force this change. No business changes its culture by simply announcing it has happened. Culture is a mixture of people's actions and internal beliefs, so while it's possible to force people to adopt systemisation practices in the short-term, changing their thinking processes takes a lot longer.

You'll have a day where it feels like every knowledgeable worker is being grumpy about the project. A manager who you thought was fully onboard bumps your training session back a month because a new client has landed. Even the owner of the business is suddenly distracted by a new opportunity for the business and starts planning for a new direction that is going to conflict with the all the hard work you've completed.

You can't be discouraged by this. Expect it. Be ready for it. And push on regardless. For systemisation to truly take hold, it must become ingrained in the very fabric of the business, becoming part of its DNA. This goes beyond simply documenting processes. Systemisation must be seen as the default way of operating.

This will only happen through consistent and persistent effort from both the leadership team and the Systems Champion. Be patient with the team but be unwavering and resolute. Remind management that SYSTEMology is a key project for the business and holding it back for short-term gain is not a good strategy. And, if you need to, have a frank conversation with the business owner about what was previously agreed upon. Remind them that changing direction isn't going to get them any closer to their goal of extracting themselves from the day-to-day of the business.

In your mind, systemisation must become a core business value for the good of the company. Everyone else needs to see systemisation as something that benefits them directly.

There's a balance to be struck here. On the one hand it's unreasonable to expect everyone in the business to prioritise SYSTEMology with the same fervour as you. On the other hand, at times you will have to be pushy to prevent the project being derailed.

This is where evidence of the power of systemisation is going to be your greatest weapon. Keep building your case. Keep strengthening your evidence. The verdict you're seeking isn't from a judge or jury. It's from every person who interacts with your organisation and can see, without a shadow of a doubt, that systems are "how things are done around here".

What success looks like

Remember the case study I shared at the end of the last chapter? The one with Shannon Smit at Smart Business Solutions?

There's another part to her story that illustrates what it will look like when you build a culture of systemisation. In a recent visit to their office, I noticed something remarkable. There was no heavy-handed enforcement of procedures, no constant reminders about following protocols. Instead, there was something far more powerful at work.

When a team member encountered a question, their instinctive response wasn't to ask a colleague or make an assumption. It was, "Is it in the system?" When someone discovered a better way to do something, they didn't keep it to themselves. They immediately moved to update the system, excited to share their improvement with the team. New team members didn't just receive an operations manual – they stepped into an environment where systems thinking was as natural as breathing.

This is the cultural standard you should be striving for. It's not about rules or compliance. It's about creating an environment where using systems becomes the path of least resistance, where working systematically feels as natural as driving on the correct side of the road. Think about that for a moment. When you're driving, you don't consciously think about staying in your lane. You don't need reminders or enforcement. It's just ... how things are done.

That's the level of cultural embedding we're aiming for with your systems.

Your Culture Evidence Portfolio

Let's gather your proof that systems are becoming "how we do things here".

Current evidence (check what you already have):

❑   A Systems Champion

❑   Checklists

❑   Documented systems

❑   Team workshops

❑   Success stories

❑   Other: ....

....

....

....

Next three pieces of evidence to build:

1 ....

....

2 ....

....

3 ....

....

18

Recruitment

IN THE EARLY 1970S, HERB Kelleher made a decision that would transform the airline industry. While other airlines were focused on rigid professionalism and technical expertise, the co-founder of Southwest Airlines had a different idea: hire for attitude, train for skill.

It was a radical notion at the time. Instead of prioritising years of aviation experience or formal qualifications, Southwest looked for people who naturally aligned with their values. They wanted team members who would bring enthusiasm, flexibility and genuine warmth to every interaction.

Their hiring process reflected this philosophy. Rather than traditional interviews focused on technical skills, Southwest created group sessions where they watched how candidates interacted and approached challenges. They weren't looking for the most experienced airline professionals. They were looking for people who would naturally thrive in their culture.

The results? While other airlines struggled with customer satisfaction and employee turnover, Southwest built one of the most distinctive and successful cultures in aviation history. They didn't achieve this by changing people – they achieved it by selecting people who were already a natural fit.

This brings me to a rather uncomfortable truth about building your systems culture ...

Perhaps the easiest way to get your team to follow systemised processes is to fire everyone who isn't immediately on board and hire new people who are motivated, accountable and love following processes.

Simple ...

Haha ...

But obviously that isn't an option ...

Or is it ...?

Haha ...

No, it's definitely not ...

But hidden in that uncomfortable joke is a profound truth. What's the easiest way to motivate people? Hire motivated people. What's the easiest way to get people to follow systems? Hire people who like following systems.

Of course you can't replace your entire team. You'll continue working with your existing staff, providing all the support, training and encouragement they need to embrace the new systems culture. But alongside this, you can make a powerful shift in how you build your team moving forward.

By being more deliberate about who you bring into your organisation moving forward, you can make the entire cultural transformation easier. Instead of constantly pushing against resistance, you can select people who will naturally align with your systematic approach.

This might not be an immediate priority if your company isn't actively hiring right now, and that's okay. This is about working with whoever handles your recruitment, whether that's a hiring manager, HR team or external recruiter, to gradually embed these principles into your hiring process.

Over time, as more systems-minded people join your team, you'll find that maintaining your systems culture becomes easier and easier. Let's explore how to make this happen, starting with some simple but powerful changes to your recruitment process.

We'll focus on a few key areas: your position description, your job ad and your interview process. Remember that your position description is an internal performance blueprint while your job ad is an external magnet for attracting the right candidates. Adding some systems magic into these three places can have profound effects.

Update your position descriptions

All good recruitment processes begin by getting crystal clear on what you're looking for. Your position description is the foundation that shapes everything that follows, from how you write your job ad to the questions you ask in interviews.

After all, how can you find the right person if you haven't clearly defined what "right" looks like?

Working with whoever handles your recruitment, you're going to weave systems thinking and systems language into your position descriptions. This isn't just about updating documents, it's about creating a clear profile of the systems-minded person you want to attract.

What might this look like in practice? Here are some examples.

Marketing coordinator

Before: Responsible for managing social media accounts, creating marketing materials and coordinating events.

This description could attract anyone with basic marketing skills. But will they thrive in your systems-driven environment?

After: Responsible for managing social media accounts using the approved social media management system and following documented content creation and posting procedures. Creates marketing materials using established templates and brand guidelines accessible for the team to follow. Coordinates events by following the documented event-planning process, which includes checklists, timelines and vendor-management procedures.

Project manager

Before: Responsible for overseeing projects, managing budgets and coordinating with team members.

This description could be from any company, in any industry. It tells candidates what they'll do but not how they'll succeed in your systems-first environment.

After: Responsible for developing and maintaining our project management software platform, ensuring all projects adhere to standardised processes for initiation, planning, execution, monitoring and closure. Uses documented templates for project proposals, status reports and risk management plans, ensuring consistent project delivery and communication. Maintains and improves standard operating procedures for:

- Project kick-off meetings and documentation
- Weekly status updates and stakeholder communications
- Resource allocation and capacity planning
- Change request handling and scope management
- Project closure and lessons learned sessions.

Notice the difference? This description isn't just listing responsibilities, it's showing candidates exactly how they'll manage projects in your organisation. It signals to systems-minded professionals that you have a mature, process-driven approach to project management.

Customer service representative

Before: Responsible for answering customer enquiries via phone and email, resolving complaints and processing returns.

At first glance, this might seem fine. After all, isn't this what customer service representatives do? But think about the kind of person this description attracts. It says nothing about how you want the work done or the systematic approach that ensures consistent, exceptional service.

After: Responsible for delivering exceptional customer experiences by following our proven customer service systems. Responds to customer inquiries via phone and email using our established customer service platform, following documented scripts and procedures for common scenarios. Makes customers feel valued while ensuring consistent service delivery through:

- Following our detailed call handling procedures and scripts
- Using our ticket management system to track and escalate issues
- Implementing our standardised complaint resolution process
- Following our documented returns workflow using our inventory management system
- Maintaining accurate customer interaction records using our CRM
- Contributing to our knowledge base by documenting new solutions.

See what's happening here? It's not just a list of tasks; it shows candidates that success in this role means embracing systematic approaches to customer service. You'll attract people who understand that great service comes from consistent processes, not just good intentions.

By crafting position descriptions this way, you're painting a clear picture of how work gets done in your systems-driven organisation rather than just listing job duties. And here's a practical tip that can make this process much easier. AI tools like ChatGPT are amazing at helping you transform traditional position descriptions into systems-focused ones. Simply feed it your existing description and ask it to rewrite it with an emphasis on systems, processes and documented procedures.

But remember, whether you're writing these descriptions yourself or getting AI assistance, the goal remains the same. You're creating a powerful filter that attracts people who will thrive in your systematic environment while gently deterring those who prefer a more freestyle approach.

Update your job ads

Your job ad is often a candidate's first glimpse into your company culture. Think of it as your opportunity to send a clear signal to systems-minded people that your organisation is where they belong.

Most job ads read like a generic wish list: "seeking motivated individual", "excellent organisational skills required", "strong attention to detail". But we're going to do something different. We're going to craft ads that speak directly to people who love structure, processes and systematic approaches.

Let me show you how to transform common job ad language into powerful magnets for systems thinkers.

Instead of writing this	Try writing this	What it signals
"Excellent organisational skills"	"Proven ability to develop and maintain efficient systems and processes"	We value systematic approaches
"Strong attention to detail"	"Meticulous approach to documentation and process improvement"	We care about quality and continuous improvement
"Ability to work independently"	"Thrives in a structured environment with clearly defined roles and responsibilities"	We have clear systems in place
"Problem-solving skills"	"Ability to analyse processes and implement systematic solutions"	We solve problems through systems
"Team player"	"Contributes to and follows team systems and processes"	We work together systematically

Start by changing the language you use. You might even include actual examples of your systems in the job ad. For instance, if you're hiring a marketing coordinator, you might write:

"In this role, you'll follow our documented content creation process, which includes our social media posting schedule, brand guidelines and quality control checklists. You can preview these systems here: [Link to relevant documentation]"

This does two brilliant things:

1. It shows candidates exactly how work gets done in your organisation.

2. It naturally attracts people who get excited about well-documented processes.

Now, I know what some hiring managers might say: "Won't this scare away candidates?" But that's exactly the point. We want to attract people who read this and think, *Finally! A company that has its act together!* while gently deterring those who might struggle in a systems-driven environment.

By incorporating these subtle yet powerful tweaks, you can transform your job adverts from generic descriptions to targeted invitations that speak directly to the heart of systems-minded individuals.

Share these examples with whoever handles your recruitment and work with them to create the perfect magnet to attract candidates that have an affinity for your systems culture.

Find more position descriptions and job ads here:

www.SystemsChampion.com/resources

Update your interview process

Now comes one of my favourite parts: the interview process. Every business recruits a little differently and there can be many stages of a recruitment process, from initial questionnaire application to multiple live interviews. It will depend on the size of your team, the role being recruited for and many other variables.

So rather than give you a very prescriptive process to follow, I want to get you thinking. The key is to add additional questions, wherever possible, into the various stages that help to reveal systems-driven candidates.

Think of these questions as your second line of defence. Your job ad has already filtered for systems-minded candidates. Now these questions, when added to your online questionnaires or within your various interviews, help ensure those who make it through truly align with your systematic approach.

I've found there are three types of questions that are particularly effective at uncovering systems-minded candidates.

Type 1: Behavioural questions

These questions reveal natural tendencies toward systemisation by exploring past experiences. They show you not just what candidates did, but how they think about processes and organisation. Here are my favourites.

- "Tell me about a time you created a checklist or system to improve your work. What prompted you to do it, and what were the results?" *Listen for: Initiative in creating structure, focus on efficiency, measurable improvements.*

- "Can you describe a situation where you had to document a complex process? How did you make it clear and easy for others to follow?" *Listen for: Attention to detail, user-focused thinking, systematic approach.*

Type 2: Hypothetical scenarios

These questions help you understand how candidates would handle situations they'll likely face in your systems-driven environment.

- "Imagine you've just inherited a process that isn't well-documented. How would you approach understanding and improving it?" *Listen for: Systematic approach to learning, improvement mindset, practical solutions.*

- "You notice team members are inconsistent in how they handle customer enquiries. How would you develop a system to ensure consistency?" *Listen for: Recognition of the need for standardisation, practical implementation ideas.*

Type 3: Direct culture questions

Finally, don't be afraid to ask straightforward questions about systems thinking. You want candidates who get excited about structure and organisation, not those who see it as a necessary evil.

- "Our company is deeply committed to systems and processes. What appeals to you about working in this kind of environment?" *Listen for: Genuine enthusiasm, previous positive experiences with systems.*

- "What do you see as the biggest benefits and challenges of working in a highly systemised environment?" *Listen for: Balanced perspective, solutions-oriented thinking.*

Remember, these questions aren't just about assessment – they're sending a clear message about your company culture. Every question reinforces that you take systems seriously, helping candidates self-select for fit. While adding a couple of extra questions won't transform your culture overnight, consistent use will gradually steer your hiring outcomes toward systems-minded candidates. You probably won't include every question I've suggested here since your team will be looking for a range of qualities, but you get the idea.

A powerful shift

By proactively weaving systemisation into your recruitment process, you're investing in the future success of your business as a systems-driven organisation. You're building a culture of efficiency, consistency and continuous improvement.

This approach attracts candidates naturally aligned with a systems-driven environment, leading to faster onboarding and greater team cohesion. It creates job adverts and insightful interview questions that act as filters, ensuring you select individuals who thrive on structure and process.

Throughout the recruitment journey, you consistently emphasise the importance of systems and prime new hires to embrace this methodology from day one. This reduces resistance to change and sets the stage for a workplace where systems are not just followed but actively championed by the team.

This shift toward a "systems are how we do things here" mentality is key to adjusting the culture. As you continue to develop this approach, I want you to fast forward in your mind's eye and try to see how the work you're doing today is going to impact the business over the next decade. The pay-off is larger than you realise.

Systems-Focused Position Description Makeover

Pick one position in your company that you're likely to hire for next:

Role: ....

Current description of key responsibilities:

1 ....

....

2 ....

....

3 ....

....

Now let's transform it to attract systems-minded people.

How will they use systems in this role?

1 ....

....

2 ....

....

3 ....

....

What documented processes will they follow?

1 ....

....

2 ....

....

3 ....

....

What systems will they help improve?

1 ....

....

2 ....

....

3 ....

....

19

Onboarding

IMAGINE STEPPING INTO AN ELEVATOR. As the doors close, you notice something odd – everyone is facing the back wall instead of the doors. What would you do?

In 1962, social psychologists conducted a fascinating experiment. They had actors enter elevators and face the wrong direction. When unsuspecting people entered, something remarkable happened. Without question or hesitation, most turned to face the same direction as everyone else.

This wasn't just about elevators. It revealed a fundamental truth about human behaviour: we naturally conform to what appears "normal" in our environment. People don't typically challenge established patterns; they adapt to them.

How can we use this to our advantage?

Most businesses struggle with getting their team to embrace systems because they're primarily trying to change existing behaviours. It's like trying to get people in an elevator to turn around when they're already facing a certain way. It's possible, but it's a challenge, that's for sure!

But what if you could set these patterns from day one?

This is where your onboarding process becomes your secret weapon. It's your opportunity to establish "normal" before any other patterns take root. When a new team member joins, they're actively looking for cues about how things are done. They want to fit in. They want to succeed. They're literally asking, "Which way should I face in this elevator?"

And yet, most businesses completely miss this opportunity. They throw new team members into the deep end with a "sink or swim" approach, or they rush through basic orientation and hope for the best. The result? It often takes months before a new team member becomes a net benefit rather than a net drain on the business.

But when you have a systemised onboarding process that showcases your company's core beliefs, something remarkable happens. That months-long learning curve can shrink to weeks or even days. Best of all, they're learning that in this elevator, everyone faces the direction of systematic thinking.

Whether your business already has an onboarding process or you're starting from scratch, you have an incredible opportunity here. Like with recruitment, your role isn't to revolutionise onboarding single-handedly – it's to work alongside whoever handles new team member orientation to weave systems thinking throughout the process.

If your business is building from the ground up, fantastic! You have a blank canvas to help create something powerful. If you're working with an existing process, even better. You can collaborate with the onboarding team to enhance what's already working, gradually embedding systems thinking at every step.

What follows isn't meant to be a complete guide to onboarding – there are already excellent resources out there covering everything from welcome packs to team introductions. Instead, we're going to focus specifically on weaving systems thinking in.

Work with your relevant team members to ensure that from day one, your new team members understand that systems are just "how things work around here".

Let's explore how to make this happen.

Day one – Welcome and orientation

The first day sets the tone for the entire onboarding experience. Focus on creating a warm, welcoming environment while conveying the importance of a systemised approach.

- **Welcome message from the founder:** Begin with a welcome video from your founder or CEO. Keep it short (3–5 minutes) but impactful. They should share the company's story and journey, bringing to life your core values and culture. This is where they paint the picture of your vision for the future, weaving in how your systematic approach helps make this vision possible. At a small business, the founder may prefer to do this welcome in person, but encourage them to at least create a video now that can be used in the future when the business has grown and they're less physically present in day-to-day operations.

- **Essential tools introduction:** Wherever possible, try to deliver their new onboarding induction through the tools they'll be using daily. It's a simple but powerful principle. Instead of just telling them about your systematic approach, demonstrate it by having them complete their first-day activities through your project management platform and get introduced to your documented processes through your systems management software. When new team members experience your systems in action from day one, they naturally begin to see them as helpful tools.

Day two – Learning the ropes

The second day can be used to address the new hire's specific role and the work they'll be carrying out. You'll quickly find this becomes so much easier once you have many of your core systems and processes documented.

- **Daily routine:** Clearly define the daily routine and expectations for the new hire, including how to log time, handle enquiries from customers and colleagues, and access support. This structure provides order and predictability and is great for helping the new team members to quickly get a sense of what they'll be doing and what is expected of them. Even though it may be a few days before they begin this daily routine, spending some time on this now will help to set the stage.

- **Review of core systems:** Guide the new hire through the core systems relevant to their role, providing hands-on training and answering questions. As with the core tools training, this is mainly about providing an overview rather than expecting instant mastery.

- **Review core values and policies:** Have them read the company or employee handbook, which emphasises the importance of systems to the company culture. (If you don't yet have a handbook, consider this a prompt to create one when the time is right. This might be a project for someone else to take on after you recommend it to senior management, but make sure you have some input into the language around this topic.) Your new hire will see how your values, policies and systems work together to create "the way things are done here". For example, if exceptional customer service is a core value, your customer service policy outlines the importance of overdelivering for customers, while your documented systems show exactly how to respond to specific customer requests using your CRM.

Day three – Deepening systemisation expectation

The third day focuses on reinforcing the systemised approach and helping the new hire appreciate that they will be expected (and empowered) to contribute to continuous improvement.

- **Systems thinking introduction:** Share your core systems philosophy with them. I love giving new team members a copy of our "Systems Thinking Video Guide". It covers fundamental beliefs like "blame the system, not the person" and "continuous improvement is everyone's job." Whether you create your own guide or use mine as a starting point (access a copy here: **www.SystemsChampion.com/resources**), the key is helping them understand that systems aren't just about following procedures – they're about creating a better way of working together.

- **Systems Champion mentorship:** Connect the new hire with a Systems Champion (this can either be you or one of your allies) for ongoing support. Ideally this champion will check in occasionally during the first few months to ensure the new hire is comfortable with the systems they're using and that they're making proper use of them. This also gives you the opportunity to track their questions, gain insights and ask for any ideas for improvement. It shows you value their fresh perspective and encourages them to think critically about their systems from day one. Another great way to reinforce systems thinking is to share five-minute case study stories showing before-and-after examples of how systems solved real problems within the business during these check-ins.

Day four – Getting hands-on

On the fourth day, I want you to focus on allowing your new hires to, wherever possible, engage in actual work as early as possible.

- **Essential, repeatable and delegable tasks:** Start with simple, practical tasks that give them a taste of real work. Find easy-to-follow processes from their department and let the new hire complete them. Assign one or two initial tasks that are well-documented and straightforward, allowing them to contribute quickly and build confidence. This is also a good opportunity to see how comfortable they are following the process steps.

Please also note, while I've shared with you some ideas for the first four days, true onboarding extends well beyond that. It's typically a 30–90 day process. I just wanted to get you thinking.

Systems for the future

Can you see how each element of onboarding becomes an opportunity to embed systems thinking? I'm guessing your mind is already buzzing with ideas specific to your business and that's exactly what should be happening. After all, you know your organisation's needs better than anyone else, including me.

Please remember, there's a lot to think about here and this sort of change will take time. Every business is different. You will have different priorities and access to different resources. Nobody's expecting this to happen overnight.

So start where you can, when you can. Pick the elements that will have the biggest impact in your organisation. Maybe it's recording that founder's welcome video, or perhaps it's documenting your core systems philosophy. Every new team member who experiences your systematic onboarding becomes another person facing the right direction, making it even easier for the next person to do the same.

Your Systems-Onboarding Enhancement Plan

Do you currently have an onboarding process? Yes / No

If no, pick three elements to implement first:

1 ....

....

2 ....

....

3 ....

....

If yes, what impression does your onboarding create about your systems culture?

....

....

....

Where are the biggest opportunities to showcase systems thinking?

....

....

....

20

System for Unfollowed Systems (SFUS)

I KNOW WHAT YOU'RE THINKING ...
This is all well and great, Dave, but you still haven't addressed the elephant in the room! What do we do with existing team members who just won't get on board?

You've created the right environment and given team members every opportunity to jump on board but they're still resisting change.

You know the ones I'm talking about. They've seen the systems. They've had the training. They understand the expectations. Yet they still insist on doing things "their way".

The challenge here is this resistance rarely announces itself openly. You won't hear someone declare, "I don't want to do this." Instead, it manifests in a thousand small acts of non-compliance, each with its own justification:

- "I've been doing this job for 15 years – I know what works best."

- "This new way takes longer."

- "I don't have the time."

- "I'll get around to it when things slow down."

- "The old way works just fine."

- "I tried it once and it didn't work for me."

But make no mistake, while these excuses might sound different, they all amount to the same thing. "I don't want to do this, I'm not going to make it a priority, and I'll keep doing things my way."

It's normal to encounter scepticism and pushback in the early days of the project, and we've covered various ways to counter this and encourage people to get on board. But if months down the line some people are still stubbornly resistant to the changes being implemented, or make efforts to derail the program, they're engaging in a posture that I call "persistent resistance".

It always plays out the same way. A team member nods along in meetings, makes the right noises about getting on board, maybe even makes a token effort for a few days. Then slowly but surely, they drift back to their old ways. Each time you bring it up, there's a new reason, a new excuse, but the underlying message remains the same: they've decided not to follow the system.

This is where many businesses get stuck. They keep accepting excuses, having "one more conversation", hoping something will finally click. Meanwhile, other team members notice. They see there are no real consequences for ignoring systems, and gradually, your systems culture starts to erode.

First, let's acknowledge something important: this isn't easy. When you've worked with someone for years, when they're good at other aspects of their role, it's natural to keep hoping they'll eventually come around. But here's what I've learned from working with hundreds of businesses. "Hope" isn't a strategy. You need a system.

Taking responsibility

Before we get into the system for identifying and handling persistent resisters, I first want to establish that tackling these individuals is not your personal responsibility. This duty lies primarily with the person's manager.

You can liaise with a manager to discuss a difficult situation, and you're going to put systems in place around this area, but when it comes to directly addressing problematic behaviour and even disciplinary action, this is outside of your authority.

Ultimately, managers hold authority over, and accountability for, team performance. So, if someone under their command isn't adhering to the systems, it's up to them to do something about it. And, depending on the size of your business, it may also be part of the HR department's processes.

And this also holds true even if a manager turns out to be a system's roadblock. In such a case, it's senior management or the business owner who needs to address the problem.

The tough conversations

The harsh reality is that a persistent resister will hurt the business and will encourage or reinforce others' efforts to do the same. So, at what point do you decide that a persistent resister is beyond help and, for the good of the company, needs to be let go? As early as possible to make an example of them? Three strikes and you're out?

The answer is, of course, more nuanced than that. And, in practice, the final decision will likely be made by the person's manager. You may, however, have some involvement in the discussion. And your experience in this matter is going to be important.

Let me be clear about what I mean by "tough conversations". We're not talking about those initial discussions to get someone back on track. Every good manager knows their first step is addressing the "I didn't know how" or "I didn't know it was my job" excuses. In fact, if you've been following along, you might have already systematically removed the possibility of using those excuses.

But here's where it gets tricky. Once you've eliminated those first two excuses, you're left with the big one: "I don't want to." And that's when these conversations become truly challenging.

These are the discussions no one wants to have, but everyone needs to know how to handle. To help navigate these waters, I've developed a framework that takes the emotion out of the equation and helps you evaluate the situation objectively. There are several key factors to consider, and before any decisions are made, it's crucial to work through each of them systematically.

The 4 C's of resistance evaluation

Conduct: How is the resistance showing up? Look at the severity and nature of the resistance. There's a world of difference between someone who occasionally slips back into old habits and someone who's actively undermining your systems initiative. Pay particular attention to how their behaviour affects team morale and performance. When one person's resistance starts infecting others, that's a red flag that needs immediate attention.

Coaching: Have you done enough? Take an honest look at the support provided. Has the "why" behind systemisation been properly explained? Have you offered adequate training and ongoing support? Has there been a clear performance management conversation outlining expectations and consequences? Do they have a clear action plan? Sometimes what looks like resistance is actually a cry for better coaching.

Commitment: Are they willing to change? Look for signs of effort, even small ones. Is the person asking questions? Seeking clarification? Making any attempts to adapt? Or are they remaining inflexibly dismissive of all systems initiatives? Is there room to further incentivise the desired action? Their level of commitment to change is often the best predictor of future success.

Cost: What's the impact on the business? Assess the broader business impact. Are there measurable financial losses from their resistance? Is it causing double work or mistakes? Is productivity suffering? Is your systems culture being undermined? Remember, the cost isn't just about dollars, it's about the toll on team morale, efficiency and your overall systems transformation.

> **PRO TIP:** I always recommend documenting your evaluation of each "C". This creates a clear record of your decision-making process and ensures you're being thorough and fair in your assessment.

When to use the SFUS

Once you've worked through the 4 C's and determined there's a genuine case of persistent resistance, you need a clear system for moving forward. Here's my four-step system that takes the emotion out of these situations and ensures fair, consistent handling of system non-compliance.

The trigger point often comes when you notice a pattern: recurring errors that trace back to ignored processes, feedback that's been given but not acted upon or systems that are consistently bypassed despite clear training and expectations. This is especially crucial when these actions start affecting team performance, customer experience or business outcomes.

Step 1: The coaching phase

This first step is crucial, and it's where many businesses rush through or skip altogether. Start by sitting down with the team member in a one-on-one setting. Your goal isn't to reprimand but to understand and collaborate. Have an open discussion about why they're finding it challenging to follow the systems. Sometimes, resistance turns into valuable feedback about processes that could be improved.

Create opportunities for them to be part of the solution. Ask them to help refine the systems they're struggling with. Document these conversations and any agreements made – this creates clarity and accountability for both parties. Look for and celebrate small improvements when you see them. Unless there's active damage being done to the business, give this phase adequate time to work.

Step 2: The performance improvement plan

If the coaching conversations aren't creating the change you need to see, it's time to formalise the process. This isn't about escalating to punishment, but about providing crystal-clear structure and expectations.

The manager should be looking to create a detailed performance improvement plan. This should outline exactly which systems need to be followed, how adherence will be measured and what success looks like. Be specific. Instead of "better system compliance", say "following the customer onboarding checklist 100 percent of the time".

Set a reasonable timeframe for improvement. I usually recommend 30 days, depending on the complexity of the systems involved. Schedule regular check-ins during this period (weekly is often ideal) to review progress, provide feedback and make any necessary adjustments.

Document everything. Every conversation, every check-in, every bit of feedback or coaching provided. You want to be sure there's no room for confusion and you have dotted all your i's and crossed your t's.

You want to give the team member every opportunity to succeed. Oftentimes this structure and clarity is enough to turn things around.

Step 3: The formal warning

When a performance improvement plan hasn't achieved the needed results, we enter more serious territory. This is where many leaders hesitate. No one enjoys this part of the process! But remember, by this point you've already invested significant time in coaching and supporting improvement. You've created clear expectations and provided resources for success. Now it's time for absolute clarity about the consequences of continued non-compliance.

Work with HR or senior management to issue a formal written warning. This document needs to spell out three things with crystal clarity: the specific systems not being followed, the impact this is having on the business and the concrete consequences of continued non-compliance. Be explicit about timeframes and expectations – this isn't the time for ambiguity.

I've found that some team members have their "wake-up call" moment here. When they see in black and white how their actions affect the business and their future within it, something clicks. That's why even at this stage you must maintain your supportive stance while being clear about the gravity of the situation.

Step 4: The final decision

Sometimes, despite your best efforts at coaching, supporting and communicating expectations, you'll reach the point where a final decision must be made. This isn't a failure but a necessary step in protecting the culture you're building. Think of it like pruning a garden: sometimes removing one resistant element allows everything else to thrive.

Before making this final decision, conduct a thorough review of all documentation from previous steps. Verify that you've provided every opportunity for improvement and met all legal and HR requirements. Consider the impact on both the individual and your wider team. Sometimes, keeping someone who consistently undermines your systems can do more damage to morale than making the tough decision to part ways.

Remember, this system exists not to punish, but to protect the culture you're working so hard to build. Used correctly, it often helps turn resistant team members into systems advocates. And in those rare cases where it doesn't, it provides a fair, clear path to necessary decisions.

> **PRO TIP:** Throughout this entire process, maintain detailed documentation of every step, conversation and decision. This both protects the business and ensures fairness to the team member. It also helps you refine this system over time as you learn what works best in your organisation.

A final note on implementation

Work with your managers to tailor this system to your organisation's needs. While I've given you the framework, the specific details (timelines, documentation requirements, meeting cadence, etc.) need to fit your company's culture and requirements. Your role is to help create this system, then step back and let managers handle the execution. Remember, you're the systems architect, not the systems police.

Case Study   Tough Call, Better Systems

When Sandra Allars founded Taking Care Mobile Massage (TCMM) nearly two decades ago, she was a solo massage therapist with a vision of bringing wellness and connection to the elderly. Today, her Melbourne-based business serves thousands of clients through home care packages, employs over 40 therapists and is preparing for a profitable exit. But the path to this success wasn't a straight line – it took making some tough decisions along the way.

The challenge: from paper to digital

Like many businesses, TCMM hit a critical moment during the COVID-19 pandemic. With staff forced to work from home, their reliance on manual processes became unsustainable. Sandra's daughter Abby stepped in as Systems Champion, documenting processes and rolling out a new suite of tools including Microsoft Teams and Asana.

While the massage therapists readily embraced the new systems, some office staff struggled with the transition. One longtime team member, in particular, became a significant challenge. Despite having the training and support available, this team member not only resisted the new processes but actively avoided them. Their resistance created what Sandra called a "black hole" in the business – a place where information flow and transparency simply disappeared.

The decision

This was a valuable team member, great at their job under the old way of working but unwilling to adapt to the new systems-driven approach. After doing her best to help with the transition, Sandra made the difficult but necessary decision to let her go and rebuild her support office team. She began specifically hiring for adaptability and tech-savviness, often focusing on younger staff members who were already comfortable with digital tools.

"It sounds terrible, but I started only hiring tech-savvy people," Sandra admits. "Most of my support staff are now under 30. They're adaptable and comfortable [with] change."

The result

The impact was dramatic. By making this tough call and ensuring everyone was following the new systems, TCMM saw fourfold growth. The business went from handling 1000 massage hours per month to targeting 2000–3000 hours – scale that would have been impossible without proper systems-driven culture and the right team.

With systems firmly in place and her team fully aligned, Sandra is now preparing for a well-planned exit from the business. Her advice to other business owners? "Having structure in your business is really important, but having the right people in the right positions is more important. That's the one thing I have learned about running a business."

Through the combined efforts of a visionary owner willing to make tough decisions and a Systems Champion who documents the process, Taking Care Mobile Massage transformed into a scalable, systematic business ready for its next chapter.

Watch the full interview here:

www.SystemsChampion.com/resources

Pillar 3 Action Items

☐ Test and validate your systems beliefs. Move beyond intellectual understanding by actively implementing systems and experiencing their impact firsthand. Look for opportunities to prove their value to yourself.

☐ Map your first followers. Identify 3–5 team members who show natural enthusiasm for systems and create a plan to support their early adoption.

☐ Create your evidence portfolio. Start collecting proof points that demonstrate how systems are becoming "how we do things here".

☐ Review your recruitment materials. Update position descriptions and job ads to attract systems-minded candidates and reflect your systems culture.

☐ Create your systems onboarding playbook. Design a structured introduction to your systems culture for new team members that sets clear expectations from day one.

☐ Build your System for Unfollowed Systems (SFUS). Document your clear process for handling system non-compliance, from coaching through to final decisions.

IMPLEMENTATION

Summary

As you approach the end of your Systems Champion journey, it's time to create an actionable plan and reflect on the transformative path ahead. Success in this role comes from enabling others and building systems that make the entire organisation thrive.

Highlights covered in this chapter include:

- The "*Matrix* moment" when you begin to see systems everywhere in your business.

- How to conduct a comprehensive systems audit across the Documentation, Tools and Culture pillars.

- Your customised 90-day implementation roadmap for systemising your business.

- Why making your team look great is the counterintuitive key to your success.

- The natural career progression that many Systems Champions experience.

- The importance of trusting the process as your business transforms.

21

Your Action Plan

DID YOU EVER WATCH *The Matrix*?[4] It's one of my favourite movies! And if you haven't yet, consider watching it as your homework assignment. There's a moment in it when the main character, Neo, experiences an awakening. For the first time, he sees the world as it truly is. Not the everyday reality he knew, but as streams of digital code. Everything that once appeared as a chaotic blur suddenly reveals its underlying patterns and structure.

You might not realise it just yet but you're going to go through a very similar transformation. At first, your business might feel like a constant whirlwind of tasks, emails and decisions. You're always playing catch-up, scrambling to put out fires. Then something shifts. You start to see the "code", the systems that underpin everything in your business. The invisible infrastructure that, when properly built, can transform chaos into harmony.

But it's not going to be easy. In *The Matrix*, Morpheus warns Neo, "You have to understand, most of these people are not ready to be unplugged.

---

4   I am so sorry for this! I know I promised no more pop culture references but it's really hard for me. I've refrained for the last eight chapters so I'm hoping you can let this one slide. This is my last opportunity before we finish.

And many of them are so hopelessly dependent on the system, that they will fight to protect it."

You'll find that some team members have become so accustomed to "the way things have always been done", even if that way is inefficient or chaotic, that they'll resist any attempt to change the status quo. They've adapted to the chaos, found their workarounds and built their comfort zones.

You'll see this show up in their excuses: "I don't know how," "I didn't know it was my job," "I don't want to," and just about every other excuse you can think of. But you know how to combat these. You know how to remove these excuses before they even become excuses.

You've learned about the three key pillars you must strengthen on the path toward systemisation bliss: Documentation, Tools and Culture.

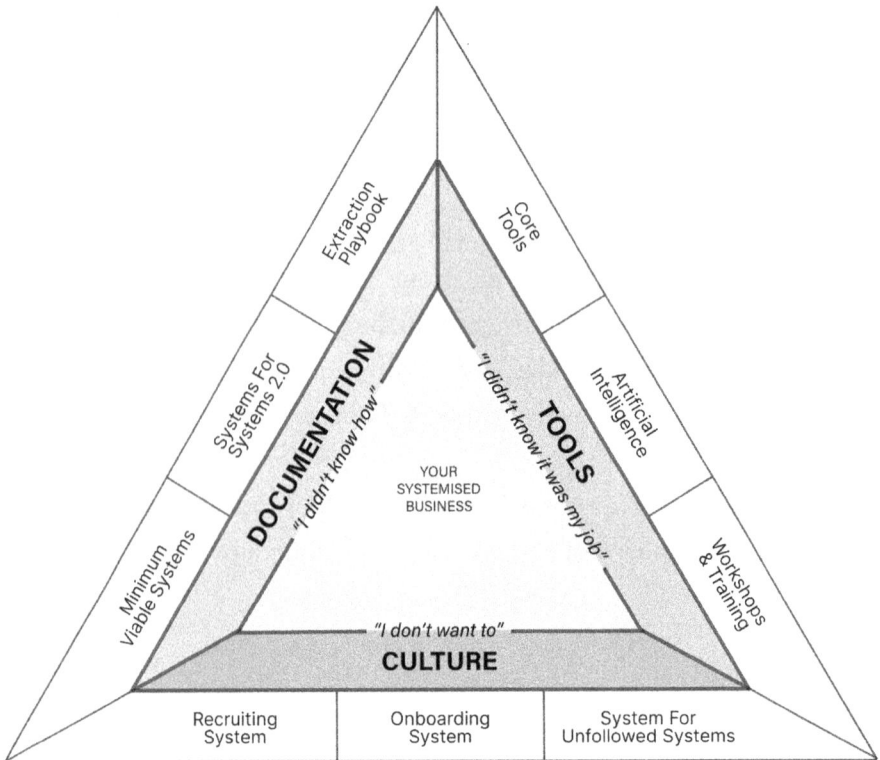

Underneath each of these areas, I have given you a range of tactics to strengthen each pillar. From using the Minimum Viable Systems (MVS) to identify the key processes, to the System for Creating Systems 2.0 to make documentation easier. From systems management and project management software, to running workshops and training to reinforce what's working. We looked at rebuilding your recruitment and onboarding processes all the way through to developing a system for how to handle when systems are not followed.

I've shared with you some of the best strategies I know. You'll build on this and you'll continue to learn and discover new strategies. You'll find new ways to solve your company's most pressing challenges. Like Neo in *The Matrix*, once you see it, you can't unsee it. Your business is a collection of systems. Every success becomes a process waiting to be documented. Every challenge becomes a system waiting to be created or refined.

We covered a lot. I know you've completed a bunch of exercises and worksheets along the way, but now it's time to pull it all together into a cohesive action plan.

You might remember earlier I mentioned the seven-stage system from my original *SYSTEMology* book (#1. Define, #2. Assign, #3. Extract, #4. Organise, #5. Integrate, #6. Scale and #7. Optimise) and how the path to systemisation isn't always as linear as we would like. In reality, every business is different and every situation is unique.

The implementation plan I'm about to share combines both what you've already learned in this book and the core stages in the SYSTEMology methodology. Use this to create your own customised roadmap that's both structured and flexible.

The SYSTEMology Implementation Plan

Step 1: Clarify your goal and pain points

Start by revisiting your notes from Chapter 5 where you described what the core systems goal meant to you and your business. (Remember? It was where you explained the phrase "The business must transition from being a business dependent on individual knowledge to one driven by documented, scalable systems" in your own words.)

Now make this more specific and practical. Identify 1–3 concrete pain points within your business that your systems work will solve. These might be bottlenecks in your delivery process, errors in client communications or inconsistencies in quality. Current pain points often provide the quickest wins and build momentum.

This approach helps secure buy-in from your team and gives you clarity before your systems audit, ensuring you'll spot the information that will have the biggest impact right from the start.

Step 2: Do a systems audit

This step is a little bit different from the ones that follow since I am going to give you more detailed direction on how to audit your current situation. In short, I want to get you chatting with your team. While the word "audit" is accurate for describing this process, I don't recommend using this word when talking to other people in the business. People associate "auditing" with tax investigations and the idea of someone combing through people's work and looking for evidence of wrongdoing. That isn't what you're doing. You're looking to work with your team to get a clear view of the current state of play.

You can simply refer to this investigation stage as "taking a systems inventory" or something similar. The goal is to discover what's already working, what needs attention and where the opportunities lie. You'll be investigating three key areas: documentation, tools/technology and culture.

Documentation

Most businesses have more systems than they realise, they're just not always obvious at first glance. Some might be formal procedure manuals sitting on a shelf, while others could be informal checklists taped to someone's monitor. Look for email templates that people use over and over, training materials tucked away in folders or even those valuable sticky note reminders that keep someone's process on track. Every piece of documentation tells part of your systems story.

Tools and technology

Next, I want you to create a master list of all the digital tools your team uses. This goes beyond just the systems management and project management tools. I want you to look into communication tools, CRM, accounting, industry-specific software, etc. You're going to have to get your head around these when it comes to documentation sooner or later, so you might as well start now.

Culture

This is where your emotional intelligence comes into play. Start identifying who in your organisation naturally gravitates toward systems and who might need more convincing. Look for those organised souls who already create their own checklists, as they're often your natural allies. Note who holds key knowledge in different areas, and who might be attached to "the way we've always done things". This isn't about judging anyone's approach – it's about understanding where you'll find support and where you might need to invest more time in bringing people along on the journey.

I'm going to be fairly broad in my recommendations, leaving you plenty of freedom to approach the specifics in whatever way you feel best suits the set-up of your business.

Start by having the business owner announce your project in an all-hands meeting. This signals that your work has top-level support and sets the tone for cooperation.

Then, schedule informal conversations with team members. Keep these casual and curious. It might be helpful to prepare a series of open-ended questions that are designed to help you understand what the person's workflows look like, what tools they use and what challenges they encounter. These questions might include:

- "Can you walk me through a typical day/week in your role?"

- "What are the most important tasks you're responsible for?"

- "Do you have any checklists, templates or documentation that you use (even if they're informal or are your own creation)?"

- "What tools or resources do you rely on to complete your work?"

- "Are there any recurring challenges or bottlenecks you encounter in your process?"

I also strongly encourage getting into the habit of recording these conversations (with permission, of course). This helps you get comfortable with recordings and the tools needed to do this. These conversations can later be transcribed and added to a searchable database so you can review the content as and when you need it.

Remember, people will have varying comfort levels with this process. Some will eagerly share their systems and ideas for improvement. Others might seem hesitant or even resistant. That's normal. Remember Morpheus's words about people being attached to their existing systems.

By the end of this inventory process, you'll have gained a comprehensive view of your business's systems landscape. You'll understand what each person in the business does and how they accomplish their tasks, even if you don't yet know every detail.

Step 3: Start with communication and quick wins

Begin running systems workshops immediately, even if they're simple (reference Chapter 16). Get your team talking about systems and the potential of AI. This early engagement sets the tone for everything that follows. Remember, you're not just implementing systems. You're building excitement about a better way of working.

Step 4: Customise your System for Creating Systems 2.0

How will you capture processes? What screen recording software will you use? How will you use AI? What tools will house your documentation? Make these decisions early – they'll form the foundation of your systems journey (reference Chapter 11).

Step 5: Focus on your Minimum Viable Systems (MVS)

Start documenting the core systems that drive your business – the ones that directly impact your ability to attract, serve and retain clients. This gives you a clear starting point and delivers immediate value to the business (reference Chapter 10).

Step 6: Spot any AI quick wins

Look for opportunities where AI can immediately reduce workload or improve efficiency with existing processes. This might be with creating content in the marketing department, drafting customer service emails, helping with data entry or streamlining other processes. Early wins here build momentum and show the power of systematic thinking (reference Chapter 15).

Step 7: Time your tools rollout

Introduce your systems management software and carefully consider when to introduce project management or other jobs management software. Sometimes it's needed immediately, but other times it's better to wait until you have core systems documented. Let your audit findings guide this decision (reference Chapter 13).

Step 8: Create your System for Unfollowed Systems

Put together your system for handling non-compliance. You want to have a clear, fair process for maintaining your systems culture. This will include briefing your leadership team to help with enforcement should the time come (reference Chapter 20).

Step 9: Adjust your people processes

Depending on your business needs, you might need to prioritise rebuilding your recruitment and onboarding processes. Remember, getting the right people and setting them up for success is crucial for long-term systems success (reference Chapters 18 & 19).

Step 10: Focus on repeatability

Once you've documented your MVS, establish a sustainable rhythm for ongoing systemisation. We do "continual systemisation sprints", which is a fancy way of saying "continuously plan out and prioritise your next batch of systems to develop". This approach prevents overwhelm while maintaining momentum. These sprints could include capturing high-value processes as they currently exist and/or re-engineering processes completely. Just ensure the systems you create add value to the business. Perfect processes aren't helpful if no one uses them.

Pulling it together

No doubt, in your mind's eye, you're starting to create your own plan. Good Systems Champions will start to do this naturally. Don't be too hard on yourself by thinking this needs to be perfect the first time round. This is a work in progress. Just give yourself enough visibility on what you think are the next couple of steps, keeping in mind some of the key milestones.

Set up your systemisation project like you would any other project. This could be done in your existing project management software if you have one, or in something as simple as a spreadsheet if you don't. The key is making it visible. Create a simple scoreboard to track your progress. List your target systems, their current status and target completion dates. This keeps you focused and makes your progress visible to the team.

Your Custom 90-Day Implementation Roadmap

Time to map out your unique systems journey below.

Days 1–30:

1 ....

....

2 ....

....

3 ....

....

4 ....

....

5 ....

....

Days 31-60:

1 ....

....

2 ....

....

3 ....

....

4 ....

....

5 ....

....

Days 61-90:

1 ....

....

2 ....

....

3 ....

....

4 ....

....

5 ....

....

Key milestones to hit:

❏ ....

❏ ....

❏ ....

❏ ....

❏ ....

Remember: Every business's journey is unique. Use the above steps and template to create a roadmap that works for you. If you'd like some extra help or a second eye, please book in for your complimentary "90-day Implementation Roadmap Review":

www.SystemsChampion.com/resources

22

Success

HERE'S A COUNTERINTUITIVE TRUTH: THE best way to succeed in this role isn't by making yourself look good, but by making your team look great! When someone follows a system you helped document and delivers outstanding results, let them take the credit. When a process improvement leads to better outcomes, celebrate the team member who suggested the change. Your greatest victories will often be invisible. They're in the problems that never happen, the mistakes that don't occur and the smooth operations that everyone takes for granted.

This might seem strange at first. After all, you're putting in tremendous effort to transform your business. But remember what we discussed about systematic change – it happens gradually, then suddenly and all at once. Your job isn't to be the hero who "fixed" everything. Your job is to be the architect who creates an environment where everyone can succeed systematically.

This is at the heart of why this role is both so challenging and so rewarding. Those early days of documenting processes, running workshops and encouraging adoption might feel like you're not making the impact you hoped for. You're putting in significant effort, but the results aren't yet visible.

Let me reassure you this is normal.

The most reliable way to transform a business is by not trying to transform everything at once. Take comfort knowing that each small improvement builds on the last. If you focus on documenting one system at a time, making each one a natural part of how work gets done, you'll find your business transforms organically as a result. Improve the whole, one system at a time.

Trust in the process

If you stick to the path and religiously work the system, I guarantee your good work will get noticed. New opportunities will naturally present themselves. Perhaps it's leading larger transformation projects, taking on operational responsibilities or even advancing into senior management. I've seen this pattern repeat itself with Systems Champions across various industries. Those who embrace this role wholeheartedly often find themselves becoming indispensable to their organisations.

Just think, you're developing rare and valuable skills. You're learning to see business operations holistically, to implement lasting changes and to lead transformation. You're becoming an expert in process improvement, team development and systematic thinking. These are capabilities that every great growing business desperately needs.

Many Systems Champions before you have parlayed this role into remarkable career paths. Some have become operations directors, others have moved into consulting and many have become trusted advisers to their business owners. By tackling this project head-on, you're positioning yourself as one of the most valuable people on the team. Someone who doesn't just solve problems but prevents them from occurring in the first place.

The beauty of this path is that it unfolds naturally. You don't need to promote yourself or campaign for recognition. The results of your work speak volumes. Your value becomes self-evident.

Your story

And while we come to the close of this book, this isn't the end of our work together. It's the beginning. I'm excited for you and I'm looking forward to hearing your story.

Throughout this book, I've shared some great case studies to provide inspiration. From Renee and Kaleb at Lime Therapy to Sandra and Abby at Taking Care Mobile Massage to Ryan and Eryn at Stannard Homes, these stories show what's possible when systems thinking takes root in an organisation.

Every business is different, but the fundamentals of successful systemisation remain the same. Each case study offers unique insights, creative solutions and practical examples of overcoming common challenges. And I trust that one day your story will be an example for others.

The good news is, you're part of something bigger now, a movement of professionals who are transforming the face of small business worldwide. We can do so much more together than we can apart!

Welcome to the SYSTEMology movement.

Sharing Is Caring

I F YOU'VE MADE IT THIS far, you're one of my people. Sadly, most people rarely finish what they start, but I can tell you're different. You're going to be one of the rare few who goes all the way.

Do you know what else I know? Rockstars like you tend to know other similar rockstars. So be sure to share this book with those who you know could benefit from it. Start with those in your organisation. Spreading the message makes your job easier. When multiple people understand these principles, you'll pick up momentum faster than you can imagine.

Beyond that, I'd like to ask another small favour to help spread the word. The reality is, most people do judge a book by its cover or at least by its reviews on Amazon. So if you've found value in this book, would you take 60 seconds to leave an honest review?

Your words might be exactly what another struggling business owner or would-be Systems Champion needs to hear. This is your opportunity to speak directly to them. Your review could quite literally change someone's life.

Leave your review here:

www.SystemsChampion.com/review

Thank you and I'll be sure to keep an eye out for your name.

Appendix

1.0 Systems Champion Position Description

Role Overview

The Systems Champion serves as the architect and guardian of a company's operational systems and processes. This pivotal role bridges the gap between business vision and day-to-day execution by creating, implementing, and maintaining the systems that enable the business to operate efficiently, consistently, and without key person dependency.

Job Summary

A Systems Champion's primary objectives include, but are not limited to:

- Advocating for and ensuring every team member follows established systems

- Developing, writing, and editing practical business systems and instructions

- Extracting and documenting knowledge from subject matter experts

- Championing the use of systems across all departments

- Training and supporting team members in system use and best practices

- Fostering a systems-thinking culture
- Continuously researching and implementing improvements in systemisation

Key Responsibilities

A. Business Operations & Strategy

- Collaborate with leadership to align systemisation efforts with business goals
- Identify operational bottlenecks and inefficiencies
- Prioritise and scope system development projects based on business needs

B. Team Development & Team Enablement

- Create systems using a variety of source materials (videos, interviews, screenshots, etc.)
- Maintain and update systems as workflows evolve
- Ensure all systems are accessible, usable, and consistently formatted
- Work across departments to ensure adequate documentation for all roles
- Train new and existing staff in the company's systems approach

C. Communication & Implementation

- Provide clear, step-by-step documentation for complex tasks
- Engage team members in the process of systems creation and improvement
- Guide the use of digital tools that support systems management (e.g., process documentation software, task management tools, cloud drives)

- Facilitate training sessions and workshops to improve systems adoption and skills

D. Systems Management

- Oversee continuous documentation and review of business processes
- Ensure all systems are captured in a designated knowledge base or systems platform
- Link documented systems to corresponding tasks in project management tools where appropriate

Expectations & Success Metrics

- Volume and quality of business-critical systems documented
- Reduction in key person dependency
- Improvement in consistency and operational efficiency
- Adoption and usage rates of systems by team members
- Measurable time and error reductions through improved systemisation

Required Skills & Attributes

A. Systems Management

- High attention to detail and strong organisational habits
- Competency with standard business software (e.g., cloud docs, spreadsheets, documentation tools)
- Ability to learn and leverage digital tools for documentation and process mapping

B. Communication Skills

- Excellent written and verbal communication
- Ability to distill complex processes into simple, actionable steps
- Skilled at interviewing team members to extract process knowledge
- Able to tailor communications for different departments and experience levels

C. Project Management & Personal Attributes

- Able to manage multiple systemisation projects with shifting priorities
- Self-starter who can work independently and proactively
- Strategic thinker who also excels in detail-oriented execution
- Embraces feedback and continuous improvement
- Adaptable to evolving tools, technologies, and business models

D. Working Conditions

- [Office/Remote/Hybrid] environment with standard office conditions
- Regular use of computer equipment and software applications
- [Insert additional working conditions relevant to your organisation]

Cultural Fit

The ideal Systems Champion is:

- Naturally organised and enjoys creating clarity from complexity
- Patient and persistent when collaborating with others

- Passionate about documentation, efficiency, and business improvement

- A collaborative influencer who leads without needing authority

Growth Opportunities

This role offers a pathway into senior operations, business improvement, or COO roles. As the business scales, the Systems Champion gains exposure to a wide array of business functions and strategic decision-making, becoming instrumental in enabling scalability and operational excellence.

Download a digital version of this position description here: **www.SystemsChampion.com/resources**

1.1 Systems Champion Position Ad

Transform Our Business Through Systems Excellence

This is a fantastic opportunity for a systems-driven, detail-oriented professional to join our team at [COMPANY]. [ADD COMPANY INFO: 1-2 sentences about what makes your company special]

About the Role

This Systems Champion position is [in person/remote], based at our headquarters in [LOCATION].

We're looking for someone who finds satisfaction in creating order from chaos and can turn complex business knowledge into clear, actionable systems. If you're energised by improving efficiency and enabling others to succeed, this role could be perfect for you.

The Opportunity

As our Systems Champion, you will be the architect of our business systems – creating, managing and optimising processes across all departments. You'll work directly with our [business owner/CEO] to implement SYSTEMology® – a proven methodology that helps businesses:

- Free up time for senior team members
- Deliver consistently excellent results to customers
- Increase capacity across the entire organisation
- Remove key person dependency

You'll establish a collaborative relationship with our leadership team to understand strategic goals, then develop the systems that help our team achieve those goals efficiently and consistently.

Your Impact

This role offers significant autonomy. You'll drive projects from start to finish, manage your own timelines, secure needed resources and deliver results without micromanagement. With this freedom comes substantial responsibility and the opportunity to make a measurable impact on our entire organisation.

What You'll Do

Business Operations & Strategy

- Partner with our [business owner/CEO] and department heads to identify critical systems that need documentation
- Map our Minimum Viable Systems and document core systems
- Identify opportunities to eliminate bottlenecks and increase operational efficiency

Systems Development

- Lead the creation of clear, usable systems using videos, screenshots, interviews and direct observation
- Continuously review and refine systems as workflows evolve
- Ensure all documentation meets high standards for quality, usability and consistency
- Create systems that enable any team member to perform necessary tasks successfully

Team Enablement

- Train team members on our systems-based approach
- Champion the adoption of systems across all departments
- Facilitate workshops to build systems-thinking capabilities

- Ensure new team members are properly onboarded to our systematic approach

Communication & Integration

- Develop clear documentation that translates complex processes into simple steps
- Connect systems with our project management tools for seamless workflow
- Foster a systems-thinking culture throughout the organisation
- Continuously improve our methodology for creating and managing systems

What You'll Bring

Essential Skills & Qualities

- Exceptional organisational abilities and attention to detail
- Strong technical aptitude with the ability to quickly learn new software
- Excellent communication skills across written, verbal and visual formats
- Ability to extract knowledge through effective questioning
- Project management experience with strong timeline management
- Self-starter mentality with minimal need for supervision
- Strategic thinking that balances big-picture vision with practical details
- Persistence and patience when working with diverse team members

Preferred Experience

- Background in operations, process improvement or related fields

- Previous experience documenting business processes

- Familiarity with project management methodologies

- Experience influencing teams without direct authority

How We'll Support Your Success

- **Strategic Partnership:** Direct access to the [business owner/CEO] to ensure systems align with business strategy

- **Team Authority:** Empowerment to work with all team members to extract, document and implement systems

- **Professional Development:** Access to world-class SYSTEMology® training and resources

- **Growth Path:** Opportunity to develop expertise across all business functions with clear advancement potential

Your Next Step

Send your CV to [NAME] at [EMAIL] with the subject line "Systems Champion Application" along with approximately 400 words on why this role interests you and what makes you an ideal candidate. Please apply by [DAY, DATE].

Download a digital version of this position ad here:
www.SystemsChampion.com/resources

2.0 Minimum Viable Systems Cheatsheet

Minimum Viable Systems™ (MVS)

1 Marketing	2 Sales	3 Finance	4 Human Resources
Lead Scoring	Adding Lead to CRM	Accounts Payable	Employee Recruiting
Lead Nurturing	Sales Playbook	Expense Reimbursement	Job Design
Lead Generation	Pipeline Management	Refund Request	Employee Onboarding
Referral System	Pipeline Forecasting	Accounts Receivable	Employee Offboarding
Content Creation	Account Planning	Cash Flow Monitoring	Leave Management
Content Calendar	Account Review	Bank Reconciliation	Daily Team Updates
SEO Optimisation	Sales Member Tracking	Petty Cash Management	Training Programmes
Social Media	Conversion Rate Tracking	Timesheet Approval	Performance Evaluation
Email Marketing	Sales KPI Monitoring	Payroll	Compensation Planning
Website Analytics	Sales Tool Optimisation	Compensation Review	Benefits Administration
PPC Ads	Sales Content Library	Financial Reporting	Compliance Processes
Social Ads	Prospecting Process	Monthly Review	Employee Relations
Retargeting Ads	Cold Calling Scripts	Quarterly Review	HR Information System
Competitor Analysis	Negotiation Strategies	Quarterly Tax Estimates	Diversity Initiatives
Campaign Reviews	Contract Review	Annual Tax Compliance	Org Design
Feedback Collection	Customer Feedback	Compliance Auditing	HR Analytics
Personalised Ads	Ongoing Sales Training	Quarterly Planning	HR Policies & Procedures
PR Outreach	Sales Software Training	Annual Budgeting	Employee Scheduling
Promotional Items	Recognition & Rewards	Budget Monitoring	Time Tracking
Influencer Partnerships	Team Building Activities	Insurance Review	Employee Feedback
Affiliate Programmes	Relationship Selling	Risk Assessment	Internal Communications
Networking Events	Continuous Feedback	Asset Tracking	Team Building Activities
Strategic Alliances	Loyalty Programme	Liability Assessment	Employee Recognition

Content Distribution
Sales Alignment
Creative Briefs
Ad Management
Affiliate Marketing
Lifecycle Marketing
Media Planning
Loyalty Programme
Seasonal Planning
Annual Plan Creation
Quarterly Strategy
Monthly Analytics
Weekly Team Meetings
Win-back Campaigns
Retention Programmes
Reputation Monitoring
Social Listening
Lifetime Value Analysis
Referral Programmes
Strategic Alliances

Repeat Strategy
Referral Strategy
Retention Surveys
Customer Segmentation
Lifetime Value Tracking
Upsells/Cross-sells
Customer Data Hygiene
Customer Education

Depreciation Scheduling
ROI Analysis
Internal Audit Prep
External Audit Prep
Software Training
Data Integrity Checks
Vendor Management
Contract Review

Resource Planning
Job Evaluation
Exit Interviews
Attendance Tracking
Training Execution
Conflict Resolution
Culture Fit Analysis

Operations

Customer Onboarding
Customer Success
Complaint Resolution
Customer Analytics
Inventory Management
Scheduling & Appointment
Maintenance Alerts
Data Entry
Simulation & Modelling
Results Analysis
Quality Assurance
Margin Optimisation

Inventory Management
Supplier Negotiation
Budgeting & Cost Control
Cost Reduction Initiatives
Productivity Improvements
Team Incentives Setting
Training Programme
Delivery Scheduling
Quality Control
Customer Feedback
Supplier Evaluation
Supply Chain Risk
Equipment Disposal

Management

Organisational Design
Scorecard Development
Monthly Performance
Quarterly Plan & Goals
Annual Planning
Issues List Maintenance
Competitive Monitoring
Budgeting Process
Financial Risk Assessment
Stakeholder Sentiment
Risk Mitigation Planning
Performance Visualisation
Strategic Goal Tracking
Leadership Training
Succession Planning
Governance Policy
Compliance Monitoring

5

6

About the Author

M Y NAME IS DAVID JENYNS, and I'm a systems devotee on a mission to free business owners worldwide from the daily operations of running their businesses. For nearly a decade, I've been pursuing this vision. I'm proud to report that my work has now impacted hundreds of thousands of business owners globally and continues to expand.

Throughout my entrepreneurial journey, I've started, systemised, scaled and successfully sold three companies. Each business became a laboratory for refining my approach. This real-world experience, along with being an adviser to hundreds of companies across diverse industries, has made SYSTEMology what it is today. Business leaders worldwide trust and reference my work because I've consistently built my reputation on under-promising and over-delivering.

My journey has come full circle, from being the business owner trapped in daily operations to now helping others find that same freedom. Today, I lead our growing community of certified SYSTEMologists, deliver workshops and keynotes around the world and host the *Business Processes Simplified* podcast. But perhaps my greatest achievement in systemisation has been creating space for what matters most: being present as a husband and father.

After all, isn't that what business freedom is truly about?

Follow my work at: **www.SYSTEMology.com**

Other Titles by the Author

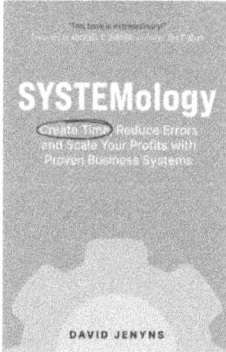

SYSTEMology: Create Time, Reduce Errors and Scale Your Profits with Proven Business Systems

SYSTEMology is the foundational book that introduced the world to "the system for systemising your business." This is where the journey began, the original methodology that Systems Champion now builds upon and evolves for our AI-driven world. Written specifically for business owners, SYSTEMology walks you through the proven seven-stage framework (Define, Assign, Extract, Organise, Integrate, Scale, Optimise) that has helped thousands transform their operations from owner-dependent chaos into reliable, profitable systems. It shows business owners that systemisation is not only achievable but gives them the confidence to see how they can make it happen.

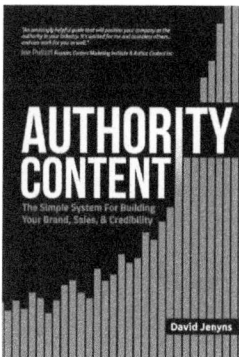

Authority Content: The Simple System for Building Your Brand, Sales, and Credibility

Authority Content is a prime example of David's long-running passion for systemisation. This first book documents his most prized marketing system built around the "3 Ps" framework (Present, Product, Promote) that he developed in his digital agency and still uses today. Authority Content shows how to systematically build credibility and visibility in your marketplace, creating a content machine that turns one day's work into months of valuable material. It's living proof that well-documented systems stand the test of time and demonstrates the systematic thinking that would later evolve into SYSTEMology and Systems Champion.

www.ingramcontent.com/pod-product-compliance
Lightning Source LLC
Chambersburg PA
CBHW030502210326
41597CB00013B/755